SpringerBriefs in Environment, Security, Development and Peace

Volume 22

Series editor

Hans Günter Brauch, Mosbach, Germany

More information about this series at http://www.springer.com/series/10357
http://www.afes-press-books.de/html/SpringerBriefs_ESDP.htm
http://www.afes-press-books.de/html/SpringerBriefs_ESDP_22.htm

Cecilia Ng
Editor

Gender Responsive and Participatory Budgeting

Imperatives for Equitable Public Expenditure

Springer

PWDC
Penang Women's Development Corporation

Editor
Cecilia Ng
Penang Women's Development Corporation
Penang
Malaysia

Licensed in Malaysia and Singapore to:
Strategic Information and Research Development Centre
No. 2 Jalan Bukit 11/2, 46200 Petaling Jaya, Selangor, Malaysia
Fax: (60) 3 7957 8343
Email: gerak@gerakbudaya.com
Website: http://sird.gerakbudaya.com
for exclusive distribution in Malaysia and Singapore.

ISSN 2193-3162 ISSN 2193-3170 (electronic)
SpringerBriefs in Environment, Security, Development and Peace
ISBN 978-3-319-24494-5 ISBN 978-3-319-24496-9 (eBook)
DOI 10.1007/978-3-319-24496-9

Library of Congress Control Number: 2015950050

Springer Cham Heidelberg New York Dordrecht London

Cover photo: Penang State and Local Government leaders with speakers and the main organising committee of the GRB Conference. Seated from left are Maimunah Mohd Sharif (President of the Sebarang Jaya Municipal Council), Patahiyah Ismail (Mayor of the Penang Island City Council), Chow Kon Yeow (Penang State Exco member and Chair of the Conference Coordinating Committee), Prof. Dr. Ramasamy (Deputy Chief Minister II), Chong Eng (State Exco member and Chair of the Penang Women's Development Corporation) and Roberta Clarke (Regional Director, UN Women Regional Office for Asia and the Pacific). The permission was granted by PWDC who provided the photo. More on this book is at: http://www.afes-press-books.de/html/SpringerBriefs_ESDP_22.htm

Copyediting: PD Dr. Hans Günter Brauch, AFES-PRESS e.V., Mosbach, Germany

Language Editing: Ms. Veena N., Bangkok, Thailand

Printed on acid-free paper

Springer International Publishing AG Switzerland is part of Springer Science+Business Media (www.springer.com)

Participants from 16 countries all over the world converged in Penang for the GRB Conference. The photo was provided by PWDC who granted permission to include it here

Conference delegates. The photo was provided by PWDC who granted permission to include it in this volume

Foreword

This important and timely book addresses the critical issue of citizen participation through a conversation between those engaged in participatory budgeting (PB) and gender responsive budgeting (GRB) approaches. The book is rich in case studies including examples from Indonesia, India, Nepal, Malaysia, Austria, Philippines and Germany. The authors, gender specialists, policymakers and academics, engage with debates surrounding the meaning of citizen participation, its practices and constraints and draw out lessons for GRB.

Government budgets reflect the priorities of those who get to influence spending and revenue-raising decisions. GRB and PB strategies offer different means of challenging traditional, and often invisible, power relationships and budgetary outcomes. Fiscal democratization through inclusive participation in budgetary debates and decision-making processes is essential if governments are to be accountable for their budgetary priorities. It also has the capacity to reveal the differential impacts on different groups of citizens. GRB and PB each emerged in the 1980s as strategies emphasizing fiscal democracy but have largely evolved separately with little overlap or sharing of lessons.

GRB recognizes that policies and their budgets often impact differently on women and men in systematic ways. GRB strategies respond to the lack of gender neutrality of budgets with analyses of the gender impacts of government spending and revenue raising. They also aim to promote actions to change policies and budget priorities in line with women's empowerment and gender equality. Framed in this way GRB requires the participation of different groups of women and men, particularly those poorly represented or marginalized. The book confronts the reality that women's participation can be overlooked in principle or be weak in practice. Sometimes an expertise/technical approach to GRB has been conceptualized as a rival to a participatory/democratizing approach. One lesson from the early GRB initiatives of Australia, Philippines, South Africa and the UK was that the two approaches were best understood as related and intertwined. The utilization and demonstration of technical expertise had an important role in the politics of facilitating women's participation and participation strategies benefited from advocacy based on expert analyses. In practice this balance has not been easy to achieve and broad participation has often been the casualty.

The economic and institutional context in which GRB operates has been critical in influencing the spaces for women's participation. For example, the Australian initiative was implemented under a reformist Labour government in the context of an expansionary Keynesian macroeconomic policy framework and a strong women's policy office within government. The women working in the specialized women's policy unit (called femocrats) were networked with the women's movement that included an active voice for women trade unionists. Feminist researchers also participated in developing concepts and tools for gender-based analyses. However, by the early 1990s the macroeconomic policy framework had become neoliberal in its orientation and anti-inflation, debt reduction, privatization and self-provision policies were adopted. In this context, many of the efforts of the GRB participants were restricted to resisting cuts in services and changes to the tax-transfer system that would adversely impact on women. When a coalition (conservative) federal government took office in the mid 1990s, the neoliberal policy framework sharpened with cuts in welfare spending and a shift in the taxation system to indirect consumption taxation. The funding of the women's policy coordinating office was slashed by nearly half. Thus a key gender policy institution fundamental to the Australian GRB initiative was severely undermined. Also, under neoliberalism, activist civil society groups, including many women's groups, were marginalized further by being positioned as special interest groups pressuring the government to allocate scarce resources away from the broader community. In the following decades, the Australian GRB initiative continued to adapt to the changing economic and institutional contexts under different governments which in turn re-shaped the space for women's participation.

This book's focus on citizen participation in budgeting is highly relevant in today's climate of austerity policies implemented in response to the 2008 global financial and economic crisis. The financial and economic crisis has resulted in the private debt of the banks and other financial institutions being transferred to governments. This once again has reminded us of the power of the finance sector and the potential limitation of states to act in the interests of less powerful groups. The budgetary responses of government have been to reduce services and benefits and cut public sector employment. The potential for women, particularly poor women, to bear a greater burden of these budgetary policies is high given their relatively greater dependence on government services, benefits and jobs. Furthermore, as the unpaid work of women is the alternative to family support services provided by governments, women's time burdens are likely to increase. The context of the economic and financial crisis and austerity policies pose many challenges for both GRB and PB including the likelihood of fewer resources inside and outside government to undertake analyses of expenditures and revenues. It also emphasizes the need for strong political mobilization to contest austerity policies for their gender, class and other biases and to bring about alternatives. GRB initiatives with strong participatory features are better placed to meet these challenges and contribute to institutional changes that foster gender equality.

The book offers the promise of a new direction for entwining women's and men's participation by building a broad-based expertise in budget analysis and good financial governance. The exemplar is the Penang Gender and Participatory Budgeting (GRPB) pilot project (2012–2014) which has been implemented at two local governments. As one who was involved in the early stages of establishing this pilot project, I am impressed with its progress over the last three years. The pilot emerged out of new political context with the election of the progressive three party coalition, called the People's Alliance. A new women's agency within the Penang state government was established and a partnership developed between academia, women's NGOs and the state government. The pilot followed a scoping exercise that recommended a model that would incorporate appropriate structures and processes for ensuring participation, capacity building, high level commitment and produce budgetary outcomes that met women's and men's needs. As the project has unfolded, it demonstrated that ordinary women's and men's participation on budgetary matters that impact on their everyday lives can be a powerful force for change.

December 2014

Rhonda Sharp
University of South Australia
Hawke Research Institute

Preface

Since 2004 there have been several Treasury call circulars from the Ministry of Finance, Malaysia, obliging government agencies at all levels to implement gender responsive budgeting (GRB). Penang is the first and only state in Malaysia to respond to this important call. While the idea to implement GRB was mooted in 2010, it was only in 2012 that a GRB pilot, with a participatory focus, was initiated by the newly established Penang Women's Development Corporation (PWDC), a state-linked women's agency. This project was implemented in smart partnership with the two local councils, the Penang Island Municipal Council[1] and the Seberang Jaya Municipal Council. Three years down the road, PWDC was keen to share and learn from the experiences of other countries, particularly on synergizing GRB and participatory budgeting (PB)—two approaches which have existed since the 1980s and have been practised all over the world. PWDC has since renamed this as a gender responsive and participatory project (GRPB).

Subsequently, a conference was organized in Penang, Malaysia during 24–25 February 2014. This Asian Regional Conference entitled 'Gender Responsive Budgeting Narratives: Transforming Institutions, Empowering Communities' was organized by PWDC in collaboration with the Penang State Government, the Penang Island Municipal Council, the Seberang Perai Municipal Council and United Nations Women (UN Women).

The conference brought together GRB and PB practitioners and experts within the Asian region and beyond to share and review experiences, reflect on successes and challenges and chart future paths that can best support the potential of both GRB and PB to strengthen each other in the budgeting process. For GRB, the question was the effectiveness of the approach in transforming institutions and to what extent it is moving processes to become more participatory. For PB, it was about the importance of integrating gender concerns into their work. As noted by Roberta Clarke, the Regional Director of the UN Women Regional Office for Asia and the Pacific in her opening address, what would be the role of GRB in opening

[1]Now Penang Island City Council.

up more democratic spaces in society? How can both GRB and PB shift our goal from mainstreaming gender in budgeting to mainstreaming gender and social justice, especially for the disadvantaged and marginalized in our midst?

Selected papers from this conference form the three themes of the book viz, the 'hybridization' of GRB and PB, the use of GRB tools for gender equality and social justice and the institutionalization of GRB and/or PB for better accountability and good governance.

I sincerely hope that this humble contribution will provide much food for thought to those of us involved in policy formulation, financial planning and better governance at all levels in society.

Penang, Malaysia
May 2015

Cecilia Ng

Acknowledgements

This book is drawn from selected papers presented at the Asian Regional Conference entitled 'Gender Responsive Budgeting Narratives: Transforming Institutions, Empowering Communities' organized in Penang, Malaysia during 24–25 February 2014. The conference itself came from the collaboration between the Penang State Government, the Penang Island Municipal Council[2], the Seberang Perai Municipal Council, the United Nations Women (UN Women) and the Penang Women's Development Corporation (PWDC), a state-linked women's agency.

I would like to thank the authors for their valuable contribution towards the fruition of this book. Thank you all for staying with us throughout this process. Heartfelt thanks go to our three reviewers whose substantive feedback helped to strengthen the various chapters. Deepest appreciation must go to Veena N., who spent many long hours, day and night, editing the various chapters for better style and presentation. It has always been an immense pleasure working with her.

A very special word of thanks goes to Rhonda Sharp, an internationally recognized expert in GRB, for graciously penning the foreword, and for being with us in our initial voyage into GRB and PB waters. Last but certainly not the least, the publication of this book would not have occurred without the strong support and encouragement from the Penang State Government, the Penang Island City Council, the Seberang Perai Municipal Council and the Penang Women's Development Corporation.

Penang, Malaysia
May 2015

Cecilia Ng

[2]Now the Penang Island City Council.

GRB: Critical Reflections—An Interactive Dialogue moderated by Prof. Dr. Rashidah Shuib (third from left) with panellists from left: Yamini Mishra (GRB Specialist for Asia-Pacific, UN Women), Dr. Elisabeth Klatzer (Austria) and Dr. Sunny George (India). The photo was provided by PWDC who granted permission to include it in this volume

Group discussion amongst Heads of Departments from the Municipal Council of Seberang Perai. The photo was provided by PWDC who granted permission to include it in this volume

Contents

Abbreviations

3Gs	Gender Equality and Good Governance Society, Penang
3R	Recycle, Reuse and Reduce
ADS	Area Development Societies
APBD	*Anggaran Pendapatan Belanja Daerah*
APE	*Anugerah Parahita Ekapraya* Award
Bappenas	National Planning and Development Agency
BFP	Budget Framework Papers
BHW	*Barangay* Health Workers
BO	Budget Office
BPFA	Beijing Platform for Action
CACs	Citizen Awareness Centres
CBMS	Community-Based Monitoring System
CBOs	Community-Based Organizations
CC	Community Centres
CDS	Community Development Societies
CEDAW	Convention on the Elimination of All Forms of Discrimination against Women
COB	Commissioner of Buildings
CPDO	City Planning and Development Office
CSOs	Civil Society Organizations
DAGs	Disadvantaged Groups
DDCs	District Development Committees
DDP	District Development Project
DPC	District Planning Committee
DPRD	*Dewan Perwakilan Rakyat Daerah,* District House of Representatives
EXCO	Executive Council
FGD	Focus Group Discussion
FKKP	Healthcare Cadre Communication Forum
GAD	Gender and Development
GAP	Gender Analysis Pathway

GB	Gender Budgeting
GBS	Gender Budget Statement
GBV	Gender-Based Violence
GESI	Gender Equality and Social Inclusion
GIS	Geographic Information System
GMS	Gender Management System
GPR	Gender-Equality Framework
GRB	Gender Responsive Budgeting
GRBC	Gender Responsive Budget Committee
GRPB	Gender Responsive and Participatory Budgeting (Chaps. 4 and 8)
GRPB	Gender Responsive Planning and Budgeting (Chap. 10)
GTP	Government Transformation Programme
HOD	Head of Department
INCLUIR	*El Presupuesto Participativo como instrumento de lucha contra la exclusión social y territorial*
IPFCs	Integrated Plan Formulation Committees
KANITA	Women's Development Research Centre
KPIs	Key Performance Indicators
KRAs	Key Result Areas
KUA	*Kebijakan Umum Anggaran*, Budgetary General Policy
LA21	Local Agenda 21
LAs	Local Authorities
LBs	Local Bodies
LGA	Local Government Act
LGCDP	Local Governance and Community Development Programme
LGU	Local Government Units
LSGA	Local Self-Governance Act
LSGIs	Local Self-Government Institutions
M&E	Monitoring and Evaluation
MCs	Master of Ceremonies
MDGs	Millennium Development Goals
MISS	Malaybalay Integrated Survey System
MMK	*Majlis Mesyuarat Kerajaan*, State Executive Council
MOF	Ministry of Finance
MoFALD	Ministry of Federal Affairs and Local Development
MOHA	Ministry of Home Affairs
MOWECP	Ministry of Women's Empowerment and Child Protection
MPPP	Penang Island Municipal Council
MPSP	Seberang Perai Municipal Council
MWFCD	Ministry of Women, Family and Community Development
NGOs	Non-governmental Organizations
NHGs	Neighbourhood Groups
NKRA	National Key Results Areas
OBB	Outcome Based Budgeting
OBT	Output Budgeting Tool

PATTIRO	*Pusat Telaah & Informasi Regional*
PB	Participatory Budgeting
PBB	Performance Based Budgeting
PBET	Participatory Budgeting Expenditure Tracking
PKPJS	Pertubuhan Komuniti PPR Jalan Sungai
PLA	Participatory Learning and Action
PRODOC	Project Document
PRS	Poverty Reduction Strategy
PWDC	Penang Women's Development Corporation
RANDA	Local Action Plan
RPJMN	National Medium-Term Development Plan
RPJPN	Long-Term National Development Plan
SIP	Soroptimist International Penang
SKPD	Local Government Working Units, Indonesia
SMCs	Supervision and Monitoring Committees
SPSS	Statistical Package for Social Science
SU	SetiaUsaha, local council Secretary
SWGs	Sector Working Groups
TAG	Technical Advisory Group
UCs	Users' Committees
UN Women	United Nations Women
UNDP	United Nations Development Programme
UNIFEM	UN Development Fund for Women
VAW	Violence against Women
VDC	Village Development Committee
VFPCK	Vegetable and Fruit Promotion Council Keralam
WBG	Women's Budget Group
WCC	Women's Centre for Change
WCFs	Ward Citizen Forums
WDOs	Women's Development Offices
WORTH	Women on the Rise Telling her Story
YDP	Yang di Pertua, President of the local council

List of Figures

List of Tables

Chapter 1
Making Public Expenditures Equitable: Gender Responsive and Participatory Budgeting

Cecilia Ng

Abstract Gender responsive budgeting (GRB) arose from the women's movement while participatory budgeting (PB) erupted from progressive circles, including the workers', slum dwellers' and peasants' movements. After about 30 years of practice, both GRB and PB are now at important junctures in their development. There have been few attempts to interlink or integrate the two approaches. The chapters in this book attempt to look at common denominators of successful GRB and PB outcomes, as well as challenges, taking into consideration the various national and local contexts. Three main themes emerge viz. the 'hybridization' of GRB and PB; the implementation of GRB tools for gender equality and social justice; and the institutionalization of GRB and/or PB for better accountability and good governance.

Keywords Gender responsive budgeting · Participatory budgeting · Synergy · Hybridization · Evolution · Democratic processes

1.1 Introduction

With about 100 countries currently implementing gender responsive budgeting (GRB), and nearly 2,800 sites practising participatory budgeting (PB), it is now not possible for policy makers and planners to avoid integrating gender and community centred concerns into their budget planning, implementation and monitoring processes.[1] While calls for more equitable sharing of resources at the national and global levels have been ongoing for a long time, these voices have become more strident and

---

Cecilia Ng, Independent Researcher and Board Member, Penang Women's Development Corporation, Penang, Malaysia. Email: cncs77@gmail.com.

---

[1]See Allegretti (this volume) for data regarding PB, and Hattam (2012) for statistics on GRB. UN Women appears to be in the forefront of GRB, supporting such initiatives in 55 countries. Within the Asian region, other international stakeholders include the Friedrich Ebert Stiftung and the Asia Foundation.

C. Ng (ed.), *Gender Responsive and Participatory Budgeting*,
SpringerBriefs in Environment, Security, Development and Peace 22,
DOI 10.1007/978-3-319-24496-9_1

better organized in the last three decades, particularly in the context of socio-economic changes and political upheavals. Indeed, it was in the decade of the 1980s, that the demand for budgets to be made accountable to the people they serve gave rise, albeit separately, to initiatives around GRB and PB. While GRB emerged from the women's movement in their call for gender equality in the context of national governments, PB's rallying cry of empowering marginalized communities to engage with local governments came from the social movement circles. In many cases, while GRB received support from national governments, PB gained traction from local authorities anxious to increase their own legitimacy as they were threatened by the growing mistrust in representative institutions as well as by uncompleted processes of decentralization, or even new recentralization strategies. Both called for participation of the people in the budget process during a period of democratic reform and change, but as the years went by, various permutations and configurations of GRB and PB evolved (Sintomer et al. 2013a; Klatzer/Stiegler 2011).

According to Sharp (2003: 9) while the goals of GRB might vary, what have remained core to GRB initiatives are threefold: firstly, to raise awareness of gender issues and impacts; secondly to make governments accountable in translating gender equality into budgetary commitments; and finally to change budgets and policies to promote gender equality. She also pointed out that participatory processes are central to the achievement of these goals.

It has been noted that in 1984 the Australian federal government was the first to develop an assessment of the impact of its budget on women and girls which was formally published as a 'Women's Budget Statement'. While the Women's Budget Statement was initiated by a Labour government with an agenda of economic and social reform, subsequent governments, both Labour and Liberal Coalition, maintained a version of the Women's Budget Statement. From the beginning feminists working within the bureaucracy, termed the 'femocrats', were central in establishing and driving the GRB initiative (Sharp/Broomhill 2013). As Sharp indicates in the Foreword this had crucial implications for the participation of the women's movement over time.

At about the same time, over in Europe, a feminist critique of macro-economic planning emerged as a result of the negative impact of structural adjustment policies and programmes, introduced in the 1980s by global financial institutions on poor countries in the global South. Women, especially from marginalized communities, had to bear the brunt of neo-liberal economic policies which reduced expenditures on health and other social related programmes, hence increasing women's unpaid work in the reproductive sphere. Feminist economists questioned the gender neutrality of macroeconomics, including the impact of budgets on gender equality and women's empowerment (see Frey 2014). A noteworthy example was the Women's Budget Group formed in 1989 in the United Kingdom, comprising feminists and representatives from civil society and trade unions to comment on and engage with the government on budget concerns. Further inspiration was provided by the South Africa Women's Budget Initiative in 1995, an outcome of the first democratic election in the country.[2] In the same year, the Beijing Platform for Action which asked governments to systematically review how women benefit from public expenditure was also instrumental in accelerating GRB all over the world—a gender equality strategy which continues to be supported by many international organizations and donors until today.

Hence, gender responsive budgeting (GRB) not only offered a formal and systematic critique of national budgets, but also provided pathways and tools on how to mainstream gender (equality) into budget planning and implementation. It encompassed political, technical and administrative processes and allowed for women's participation in various stages of the budget cycle. However there is no single blueprint in the GRB strategy, as these initiatives vary from country to country depending on their scope, nature of the political regime/institution, the actors involved and their format, to name but a few characteristics. Klatzer/Stiegler (2011) note a number of such approaches ranging from mainstreaming gender in the budget process to analysing pilot programmes, and implementing GRB in combination with PB approaches, especially at the local level.

On the other hand, PB emerged in the late 1980s out of a different impetus. PB is known to have started in Porto Alegre, Brazil—its cradle so to speak, at a time of democratic struggles and regime change. There was a crisis in legitimation of previous repressive states whose rule was challenged by the new social movements of the day which demanded an end to corruption as well as measures to counter neo-liberalism. The election of a left wing government with the support of civil society, particularly the labour, slum dwellers and landless peasant movements, led then to the birth of PB—a tool whereby citizens have a right to decide how public resources are to be utilized. PB 'à la Porto Alegre' thus involved both top-down and bottom-up processes where genuine mass public deliberations took place from neighbourhoods to districts right up to the city level.[3]

Accordingly, PB as an exciting innovation in local democratic reform and good governance has three clear underlying principles, viz. grassroots democracy, social justice and citizen control. While defined as the 'participation of non-elected citizens in the conception and/or allocation of public finances', five other criteria are important in distinguishing this domain of practice. These include: (1) discussion of the financial and/or budgetary dimension; (2) involvement of the municipal council; (3) repetition of the process; (4) presence of some form of public deliberation; and (5) accountability to output (Sintomer et al. 2013a). As with GRB, there is no

[2]The institutional bases for participation for the UK and the South African Women's Budget initiative were very different from those of Australia. Nongovernmental organizations, researchers and academics provided the assessment of gender impacts in the UK and South Africa, rather than the bureaucracy. Over in Asia, the Philippines Gender and Development (GAD) Budget Policy, initiated in 1995, mandated that all government agencies set aside at least 5 % of their total budget for gender and development (Illo 2010). I would like to thank Rhonda Sharp for her inputs into this part of the introduction.

[3]The section on PB relies on the research and incisive writings of Allegretti (2014), Sintomer et al. (2008, 2013a, b).

definitive blue print, as the original idea of PB has now become more widespread with many models and hybrids around the world.

Indeed, there have been various interpretations of 'participation' in the PB context. A useful unpacking of the notion is the idea of 'participation by invitation' and 'participation by irruption'. The former is where the government seeks the participation of different social groups of women and men, often a top-down creation; while the latter is the situation where the capacity for self-mobilization is achieved from within, thereby leading to the 'transformation of institutionalized spaces of participation' (Allegretti this volume). The recent PB world report provided six typologies of PB, viz. participatory democracy, proximity democracy, participatory modernization, multi-stakeholder participation, neo-corporatism and community development (Sintomer et al. 2013a). While there aren't any simplistic cut and paste formulas of previous experiences, successful PB experiments have been a result of political will of public institutions, self-organizing capacities of the communities involved, proper organizational design and financial autonomy granted to the institutions promoting the process (Allegretti 2014).

After about 30 years of practice, both GRB and PB are now at important junctures in their development. How does one measure the impact of the myriad variations of both initiatives all over the world? Has GRB now become a mere technocratic counting exercise (Mishra/Sinha 2012; Frey 2008)? Has PB become routine and merely symbolic (Sintomer et al. 2013b: 21)? Is GRB more suitable for the national level, and is PB more suited to local government and local negotiations? How can each approach lend more strength and impact each other's dreams and goals?

The main questions are, as always, what has been achieved, and how is this measured? Who/which groups participate and benefit? While important, this book does not pretend to completely answer these big questions. It focuses on GRB and interrogates the nature of participation in its conceptualization and implementation in the various country case studies which have tried to amalgamate both models. These include examining the structures, mechanisms and processes which have been put in place to ensure that public expenditures are more need-based, equitable and accountable. The chapters in this book attempt to look at common denominators of successful GRB and PB outcomes, as well as challenges, taking into consideration the various national contexts and multiple identities in society. The fundamental questions this book seeks to address are firstly, what are the challenges to institutionalization and participation in GRB? Secondly, can GRB, a statist project, be participatory in the 'true' sense of the word? And last but not least, what have—or have not—been the synergies between GRB and PB in terms of evolving into newer types of innovation in policy and budget planning?

Three main themes are identified viz. the 'hybridization' of GRB and PB, to borrow from Allegretti (2014), as narratives of change and innovation; the implementation of GRB tools for gender equality and social justice; and the institutionalization of GRB and/or PB for better accountability and good governance. Four papers constitute the first theme, while the second and third themes comprise two and three chapters respectively.

1.2 GRB and PB: Narratives of Change and Innovation

The four chapters in this section query whether there can be any meaningful relationship between GRB and PB. Central to the issue is how one defines 'participation', if indeed there is a call for GRB to be more participatory. The first two chapters, from experts who are also practitioners in their respective fields, are discursive, while the next two chapters provide concrete case studies in Penang (Malaysia) and Kerala (India) to evidence (or not) the GRB-PB synergy.

Regina Frey makes a provocative and incisive point that 'PB is not automatically gendered and GRB is not automatically participatory'. In fact, she candidly calls out that GRB was not originally meant to be participatory, since the intention was to engage with national budgets, compared to PB which had origins in a grassroots perspective. That said, Frey asserts that GRB and PB "can and should complement each other given the clear commonality between both strategies. However challenges and limitations need to be recognised". She begins by pointing out that participatory approaches have been criticized for being gender naïve and gender blind, an argument also articulated by Giovanni Allegretti. In fact, even if gender was included as a concern in some PB experiments, it was done in a superficial manner in terms of the counting of numbers as the focus was on access to, and not on control over, resources. Thus what is important in the context of participation is the transformation of power relations which includes dismantling hierarchies based on gender, caste or race and ethnicity. In order to ensure that such hierarchies are not reinforced, which she noted has happened in some German PB examples, she proffered recommendations to 'stabilize' these unjust power relations, one of which was to include gender as a structural category in PB, a point also taken up in Chap. 3.

Frey is equally critical of GRB as she states that GRB does not often refer to participation but focuses on 'technical procedures and expert budgetary knowledge', including the popularly referenced seven tools of gender budget analysis. But she notes that it is possible to include participatory approaches in these tools and provides various examples of entry points for participating in the budget cycle. Using German experiences, the chapter demonstrates ways, possibilities and limits to interlink participation and gender equality given the assumption that the main objectives of GRB are better and more effective distribution of resources, greater transparency and the empowerment of disempowered social groups.

In Chap. 3, Giovanni Allegretti and Roberto Falanga are similarly scathing when they quote the first comparative research of European PB experiences which clearly argued that "PB almost never contributes to changing the social roles of men and women … despite the claim that almost everywhere, women appear to be involved in them in a considerable way". Like Frey, they assert that when gender becomes a 'variable', the focus has been on the quantitative aspect of women's presence in the participatory processes, thus avoiding issues of gendered power relations and the equal valorization of women's voices. Mere numerical presence gives the illusion of equality, and power relations are not questioned, and even deliberately sidestepped.

Allegretti and Falanga provide some deep insights regarding the weak commitment of PB to gender issues, lessons which would be useful for all stakeholders to contemplate and rectify. This, they state, ends up diluting PB. But this is not only the weak link in PB as the problem, they add, also lies in feminist studies and their focus on descriptive rather than substantive representation—an ongoing debate for some years now (Tan 2011; Philipps 1995).

Arguing that without such inclusion and without coherent measures of affirmative action PB would not obtain the desired social inclusion effects, Allegretti and Falanga present interesting examples where gender mainstreaming was or was not a focal point of the PB process design. The chapter highlights the diversity of measures needed to involve and empower women, particularly in the context of a male-dominated culture. Three success stories—Cotacachi in Ecuador, Rosario in Argentina and New York in the United Sates—are showcased to show the inclusivity of the PB-GRB articulation as they embrace at the intersections of gender, economic status, race and age. According to Allegretti and Falanga, the important lessons to be learnt are that there should be explicit goals and specific tools, including specific and coherent measures to overcome these multiple layers of exclusion. If not, both PB and GRB would remain a simplified, romantic notion and turn into an(other) unfulfilled promise. They argue that only the capacity to analyse and control all budget allocations in public policies will allow us to really understand how much public spending can reinforce or reduce gender asymmetry and power relations. This reflection is illuminating as the majority of PBs around the world today limit their transparent management only to the small parts of the budget which are devoted to the participatory process, leaving all the rest in the shade.

The issues and debates discussed above are further interrogated in the next two chapters by Shariza Kamarudin and Sunny George, presenting concrete GRB-PB experiences as well as challenges on the ground. Concerns include to what extent one has ensured that the voices of women and men, particularly from disadvantaged groups, are heard. Can one confine one's work just to the budgeting process, or does the empowerment of women and men in communities imply a more major transformation? The next two chapters critically share the nature and extent of community participation and empowerment in the context of state and local government budget structures and processes.

In the fourth chapter, Shariza captures in detail the Gender Responsive and Participatory Budgeting (GRPB) pilot in the state of Penang, Malaysia. Launched in 2012 as a flagship project of the Penang Women's Development Corporation, the project in partnership with the two existing local councils was an exciting experiment in attempting to interface, if not amalgamate, both GRB and PB models. Indeed, this was only possible with the change in government in 2008, an opposition regime which extolled good governance, accountability and transparency, including in the arena of public expenditure. Women's NGOs and academics utilized this window of opportunity to insert gender justice into the state agenda as the newly minted politicians were more amenable to both GRB and PB strategies. At the same time, the local government was also being modernized with plans to transform its public delivery system and to be closer to its clients—the residents they serve. A happy political and administrative conjuncture of sorts provided the pathway to implement this innovative pilot for the first time in Malaysia, begging the question: Is a 'particular' type of government an important factor for the successful combination of GRB and PB?

The chapter studies the community-based component of the pilot in two council-owned and -managed low cost apartment complexes. It examines the process and potential for dialogical action in both GRB and PB and argues that all engagements with communities must develop critical frameworks that explore whether diverse and fractured communities can develop agency, organize and transform themselves. A unique four-phase methodology was developed and implemented from 2012 to 2014 to ensure the participation of various groups—elderly, adults, young, disabled—in the two housing complexes. The completion of Phase 4 was a learning experience in interactive and citizenship participation whereby the GRPB methodology "gave the people the power to play their role as citizens and to become partners—not just beneficiaries—together with the local councils in formulating policies and making decisions in critical areas that affect their lives". Women were also empowered in the process as they were the movers in their own communities, with some of them forming and registering an autonomous women's group in one low cost apartment. Could this be the unique combination of 'PB by invitation and PB by irruption' discussed earlier? At the same time, as with most pilots, sustainability is always a concern as GRPB processes need to be further negotiated, fragmented communities need to be in solidarity with each other, women's leadership needs to be further enhanced and GRPB needs to be effectively mainstreamed in the planning, budget and administrative policies of the state.

The fifth chapter by Sunny George delves into the experience of one local government in Kerala, India, a state well-known for its democratic decentralization. It discusses how participatory planning and budgeting works in the context of a significant transfer of power to the local governments and from thence to local bodies, a model existing well before PB was created in Brazil. Gender concerns were included through a women's quota (one third) for representation in local bodies, reservation of 50 % of seats in local governing bodies for women, in addition to 10 % of the total development budget earmarked for the Women's Component Plan. Besides setting in place the important quotas, another critical pre-condition was the explicit commitment of the state government to the principles of good governance, viz. autonomy, subsidiarity, complementarity, uniformity, participation of the people and transparency and accountability.

In addition, the establishment of clear structures and mechanisms as reflected and practised via the People's Plan, paved the way for a bottom-up planning process—starting at the community level all the way to the district/municipal levels. Through this participatory process, the aspirations of the people are formulated into implementable project proposals by Working Groups, which are then consolidated through a series of further discussions. George provides interesting learning experiences from the approach used by *Kudumbashree*, a state-based multi-faceted approach for poverty reduction and women's empowerment. Other writings (Mishra 2011) have also attested to how Kerala is forging new frontiers in GRB whereby it becomes a transformative tool for improved public policy, which is then translated into and responds to the inequalities faced by women, especially those at the grassroots. The chapter concludes with the identification of challenges and elements of success in participatory planning and budgeting, several of which are also echoed in the ensuing chapters.

1.3 Implementing GRB Tools for Gender Equality and Social Justice

Over the years, GRB has developed a systematic series of approaches and tools to mainstream gender into the budget process. Admittedly, the most famous is the set of seven tools developed for the Commonwealth Secretariat detailing how to 'do' gender budgeting (Elson 1996; Budlender et al. 1998; Quinn 2009). Countries have then developed and adapted this methodology based on their specific country experiences (Ministry of Women, Family and Community Development and the United Nations Development Programme 2005). In the same league is the popular five step approach on how to conduct gender budget analysis as well as the three-way budget (on expenditure) categorization,[4] all useful tools which have been utilized by many countries (see Nepal, this volume). Not least is the importance of sex-disaggregated data and its potential role in supporting gender equality and the processes of transformation and empowerment. While the collection and use of data cannot be underestimated as it enables actual tracking of who benefits, thereby highlighting gender gaps, there have also been critiques that a focus on merely 'sex-counting' could reduce GRB to a mere technical exercise and distract planners from the real macro issues at hand (Jhamb et al. 2013). As Frey (2008: 8) argues, 'while the administration is busily counting at the micro level, decisions are taken at the macro level which may even contradict gender equality'.

The chapters by Elisabeth Klatzer and Herculano Ronolo counter such critiques with hands on and practical examples on how to integrate gender equality into budget processes and into a holistic data system. Gaps can not only be identified, but budgets can then be decreased or increased accordingly to bridge these gaps.

Klatzer begins Chap. 6 with a clear assertion that if fully implemented GRB inherently implies participatory approaches to budgeting, and 'can be even more effective if gender equality and women's rights advocates from within and outside

[4]See UNIFEM (2000: 116) for an elaboration of the three-way categorization espoused by Rhonda Sharp, which divides expenditure into gender-targeted, equal opportunity for civil servants and mainstream expenditure. Of late Banerjee has proposed three categories on public schemes which comprise relief policies, gender reinforcing assistance and empowering schemes (Mishra/Sinha 2012). Another well-known tool is the five steps towards a gender sensitive budget, the latter of which was developed by Debbie Budlender together with the South African Women's Budget Initiative and Gender and Education and Training Network (Quinn 2009).

the public administration are closely involved in these processes'. The chapter highlights how GRB can be integrated into the four stages of the budget cycle—budget formulation, approval, implementation and control. Klatzer argues that each of these stages offers possibilities for further initiatives and actions for both government as well as non-government actors, presenting different GRB instruments and tools for use at the different stages of the budget cycle. What makes this chapter very useful are the examples of actual GRB work at different stages of the budget cycle, taken from case studies around the world.

The example from Austria is attractive as GRB has been legally enshrined into the Austrian Constitution as of 1 January 2009 through Article 13(3) which states that: 'Central government, the Lander and local authorities must strive for actual equality of women and men in budgetary management' (Klatzer/Stiegler 2011:5). In addition, the country has also shifted to performance based budgeting (PBB), also known as outcome based budgeting (OBB) where the budget circular has included instructions to achieve gender equality in PBB by 2013. The chapter relates how in the context of outcome objectives, each ministry had to submit five such objectives, with at least one of them related to gender. Subsequently, each of these objectives was to be accompanied by corresponding activities, indicators and time lines, deepening our understanding of the GRB processes. To be sure, the chapter reveals the complexity of mainstreaming gender (equality) in the budget process and why it is a tremendous challenge to key stakeholders, as it not only demands gender and expert knowledge but also strategic coordination, and most of all political will to ensure that change towards gender equality, participation and women's rights is ensured and delivered.

In Chap. 7, Ronolo presents the Malaybalay Integrated Survey System (MISS) and the experiences of the city government of Malaybalay, Bukidnon, Philippines. MISS is a comprehensive and institutionalized data gathering process whose goal is not only to improve its programme and service delivery, but also to improve governance and promote accountability in resource allocation in line with the objectives of the national government. In addition, MISS has also been used as a tool for gender responsive budgeting and has made the local government unit more results oriented in relation to the needs of its diverse population. This chapter presents a detailed account of the basis of the survey tool, from the questions it poses to the enumerators who conduct the survey, the coverage and the reasons why it is considered an empowering and competence-building process. As an empowering tool, MISS has succeeded in providing local *barangay* health workers, almost all of them women, with the requisite competencies not only in keying in and coding the survey data, but also in voicing community concerns to the city council almost instantaneously. They can now participate in and use the data to plan for their *barangay*; and poor rural women, who were left out of the system earlier, can now be counted. The data, for example, has shown that women need agricultural training in their own right (and not as stand-ins for their husbands), more health services and child care support. The system has also been useful in identifying programmes and activities in line with the mandated 1996 Gender and Development (GAD) budget in which 5 % of the total budget of every government related agency is allocated to gender concerns, paving the way for GRB. Most importantly, Ronolo notes that the integration of MISS into the Geographical Information System (GIS) has also created opportunities for the interconnection between qualitative analysis of the situation with the quantitative data of MISS.

1.4 Institutionalizing GRB: Towards Better Accountability and Good Governance

The third and last section is a conversation about the challenges of institutionalization of GRB, and of budget accountability. A significant objective of GRB is to persuade respective governments to integrate GRB processes and tools into their workings as part of gender mainstreaming. This is the process of institutionalization. The other important task is that, as one of the new tools of vertical accountability, GRB has also emerged as 'an effective budget transparency and accountability tool' in its quest to contribute to deepening democracy and empowering the marginalized in society.[5] This section queries the extent to which GRB has been successfully institutionalized at national and sub-national levels. What have been the crucial components of success? Again, GRB initiatives are merely weak strategies if not institutionalized permanently and accompanied by adequate and strong monitoring and auditing mechanisms.[6] The next three chapters demonstrate three different modes and methods of working with elected representatives and government servants and examine the sort of structures and mechanisms that have helped to establish GRB within the relevant institution(s). While the chapters by Aloyah Bakar, Patahiyah Ismail and Maimunah Mohd Sharif (Chap. 8) and Purusottam Nepal (Chap. 9) are narratives about the state (Malaysia and Nepal) and its institutions in relation to GRB, the final Chap. 10 by Agus Salim discusses how civil society organizations engage with government as implementing, advocacy and monitoring partners in realizing GRB in Indonesia.

Aloyah, Patahiyah and Maimunah centre their discussion on the challenges of institutionalization of GRB initiatives with a participatory emphasis, in two local councils in the state of Penang, Malaysia. It specifically analyses the readiness of the two local authorities in Penang to institutionalize GRB within their respective organizational milieus. Using Kelleher and Rao's framework on organizational

[5]See the draft concept note and discussions in the UN Women Asia Pacific Consultation on "Making Budgets Accountable to Women", 30 September–1 October, 2014 held in Bali, Indonesia.

[6]The example from Malaysia is illustrative. A pilot on GRB comprising five ministries began in 2003 with the support of UNDP. However, since then there has been little follow up. Through the years, various Treasury Call Circulars have obliged government agencies to implement GRB through a gender and social analysis (via sex-disaggregated data for example) of their respective programmes. To date, there has been no report of such obligatory analysis.

transformation, they ask the following questions with regards to GRB. These are: (1) Is there a women's constituency advocating gender issues that will be taken up by the organization? (2) Are these external advocates able to negotiate with those wielding power in the organization? (3) To what extent is there a culture of openness, dialogue and understanding for new directions? (4) Are sufficient resources and knowledgeable people available to lead the change? and (5) Have new methodologies been developed to ensure that gender equality efforts are supported by those within the organization?

While not linear, several of the above spheres of change, in hindsight, were actually taken up in the process of institutionalizing the GRB pilot in Penang. For example, lobbying and sensitizing policy makers and political leaders by women academics and gender advocates occurred in the formulation and pre-establishment stage of the project. Negotiations with the powers-that-be resulted in local authorities and influential political leaders playing vital roles in creating an enabling and supportive environment to make GRB a reality in their respective contexts. All the same, any innovative idea would need to be accepted by personnel within the organization. The chapter points out how commitment and political will notwithstanding, competing priorities and lack of knowledge, among others, often stand in the way of GRB implementation. To overcome this, new structures had to be created and organizational processes had to be re-invented. Finally, the visibly positive results at the community level through peoples' participation were instrumental to organizational buy-in by local government (see Shariza, this volume). The way forward, Chap. 8 suggests, is both a participatory approach as well as a legal commitment to GRB structures and processes—to ensure successful transformation of the mainstream towards gender inclusivity and sustainability of people-centred governance.

Formally introduced in Nepal in 2007/2008 through the initiative of the Ministry of Finance, the country has been hailed as a good practice model for GRB in the Asia-Pacific region. Nepal in Chap. 9 details the important changes in structures and processes of budget reform at the national level that have allowed for a holistic GRB process. These include the introduction of a gender code classification system for programmes and projects; institutionalization of a GRB committee within the Ministry of Finance; use of gender and social inclusion disaggregated indicators; and an outcome monitoring mechanism. As in Kerala, the mandatory policy of 33 % women's representation in various grassroots institutions and a mandatory provision of 10 % of allocations to women's programmes from the programme budget at local level helped to accelerate the institutionalization of GRB at national and sub-national levels.

What is noteworthy is the classification and combination of quantitative and qualitative methods to evaluate the responsiveness of government programmes, two of which include the participation of women in programme formulation and implementation while the other is in relation to the reduction in workload through a qualitative improvement in their time use. Nepal provides concrete examples on how these funds were, or were not, utilized effectively—all important learning experiences in the equitable and proper/accountable allocation of funds. Also of interest is the experience of the Local Governance and Community Development Programme (LGCDP) which attempts to promote gender-sensitive participatory and decentralized governance. Nepal notes that in the first phase (2008–2012), the LGCDP created some 40,000 grassroots institutions with about 50 % women membership, a laudable achievement indeed. In addition, social accountability mechanisms, such as public hearings and community-based monitoring, were established for citizens to track government budgets. The chapter closes by acknowledging that challenges still exist as to the transformation process at the institutional level; for example, the impact of cultural norms, practices and gender stereotypes often impede progress towards achieving gender equality. Nevertheless, the chapter ends on a positive note as the next phase (2013–2017) has put forward a framework focusing on citizen's empowerment for better participation in local governance and the budget process.

Similarly, in Chap. 10, Agus documents Indonesia's journey of a gender mainstreaming policy issued by a Presidential Instruction, to integrate gender concerns into all development programmes, from planning to budgeting and the monitoring and evaluation of these activities. All this was possible within the context of democratic reform and the push towards sub-national autonomy. Since then, systematic regulatory frameworks and mechanisms have been developed by key (and influential) agencies, including the Ministry of Finance and the National Planning and Development Agency in collaboration with the Ministry of Women as well as civil society. At the local level, this has been followed up by initiatives from the Ministry of Home Affairs via the issuance of policy guidelines for provincial and district level administrations to integrate gender more effectively into their planning, budgeting, monitoring and evaluation processes. Concomitant interlocking and sometimes overlapping structures have been put in place.[7] All in all, it shows that political will and commitment, although top-down, are crucial in ensuring that Gender Responsive Planning and Budgeting (GRBP) as it is known in Indonesia, is taken seriously by policy makers and implementers.

The unique mode of GRPB implementation undertaken simultaneously at the national and sub-national levels by government and/or by civil society, with the support of international donors, is especially fascinating. Agus presents some case studies, particularly from PATTIRO, his own organization, to highlight how, in several areas at the sub-national level, civil society played a vital role in institutionalizing and monitoring GRPB. Policies to promote transparency have been effected by the actions of local groups which have also been able to influence budget allocations. With the opening up of democratic spaces in Indonesia since 1998, the role of civil society, at least in the arena of GRPB, has been re-defined from critiquing and lobbying the state to a more nuanced position of critical engagement. Strategies used by civil society organizations to promote GRPB

[7]Satriyo (2014) refers to the sometimes overlapping, inconsistent and poorly implemented laws and regulations of the GRB policy environment in which women's and civil society groups operate.

implementation include partnership arrangements, facilitation of its policies on the ground and monitoring and advocacy to ensure that these policies are implemented. At the same time, CSOs maintain a critical view of the various GRPB policies, a stance of being 'in and out of the state' as such.[8]

1.5 Gender and Participatory Budgeting: Towards a New Architecture?

In attempting to interface and synergize GRB and PB, the chapters in the book foreground several features common to both approaches. First, it is interesting, but perhaps not surprising, that both strategies emerged within a particular historical context of social and political change. Nonetheless, the drivers in this period of new social movements were different and moved separately; GRB arose from feminists and the women's movement while PB erupted from left wing circles, including the workers', slum dwellers' and peasants' movements. They were well accepted by political regimes, one at the national level, the other at the local, mainly as proponents of democratic change and good governance. Participation, with all its diverse interpretations, was also important although GRB was more 'by invitation' involving administrative and technical processes, compared to PB which was grassroots based with the intention of citizen control of resources and budget allocations.

Today, both approaches are still journeying separately with their various permutations through the years. However, the chapters in the book depict the multiple terrains and various attempts which have been taken to ensure that GRB consciously adopts PB methods while PB cases also integrate gender equality and women's empowerment into their advocacy structures and processes. At the same time, several concerns still need to be addressed if this new architecture is to take root.

An important concern is the meaning if not operationalization of 'gender'. The debate on women's quotas continues in the GRB and PB discourse as the chapters enquire whether simplistic sex counting in gender budget statements is sufficient to ensure gender equality and women's empowerment. The various authors are clear that while quotas and number crunching are important as they make visible the gaps in the distribution of resources, it is necessary to go beyond mere numbers. In fact, the notion of inclusion means not excluding other hierarchies and identities based on age, ethnicity and space, among others, so as to not reinforce or reproduce existing power relations as elucidated in several examples in the book. Allegretti correctly argues that GRP and PB must be ready to introduce a multi-layered

[8]See Rai (2000) and the chapter entitled "An Unholy Alliance? Women Engaging with the State" in: Ng et al. (2006) on the uneasy relations between state and civil society, especially the women's movement.

approach to 'plural and divergent exclusions' as gender inequalities create different conditions for participation. It needs reminding that GRB and PB are thus political projects, aimed at the transformation of power relations within the ambit of good governance. If not, as many of the authors warn, they might just be superficial show pieces boasting their successes under the illusion, nay romance, of change.

But the future is bright. Despite the challenges and limitations of this inter-linkage, the chapters reveal that such dynamic innovation is possible if several salient factors are in place. Political will and commitment within democratic settings are absolutely necessary as pre-conditions to successful GRB-PB partnerships in whichever context they are located. Other factors include quotas in terms of women's representation and budget commitments as the former ensures that gender is 'structured' in the equation with clear goals and explicit measures, while the latter ensures sustainability. Another important item is the existence of a clear regulatory framework with concomitant mechanisms to ensure participation and accountability. The state is a site of power; thus in the absence of appropriate mechanisms, inter and intra power contestation will occur among the different stakeholders along the budget cycle. And last but not least is the nature of peoples' participation, which has to be constantly critiqued and re-articulated as social group dynamics change and as regimes shift in today's fast changing global economy.

As countries move from dictatorial and elected-elite democracy towards, hopefully, more genuine grassroots democracy, we see opportunities and spaces for increased participation and good governance opening up around the world. The time is ripe for a new synergy of gender responsive and participatory budgeting to make public expenditures more accountable to the people. Certainly, a profound implementation of GRB will not be possible without opening budget processes for participation at all stages.

Besides the above, what are the other concerns in the future? This book focuses on expenditures and not on revenues and is thus limited in its analysis and presentation. Nor do the chapters examine the care economy, important considerations in budget discussions.[9] A major challenge to all governments today is the economic crisis giving rise to insecurities which are then translated into austerity budget cuts, especially in the social sectors which impact women, particularly the poor, more than men (Hattam 2012). GRPB advocates have to be able to engage at both the macro, meso and micro levels to make governments politically accountable to the people and their multiple identities, if not exclusions, which after all is a principal consideration of both approaches. In adopting the best of GRB and PB, there is hope for a more democratic and just future.

---

[9]It has been noted that the UK Women's Budget Group focused originally on taxation and state transfer of funds. Uncovering the gender bias of austerity policies has become a focus in the context of the economic crisis (Klatzer et al. 2011; Grown 2005). The relationship between the care economy and budgets has been brought up since the 1990s (see Elson 1996).

References

Allegretti, Giovanni, 2014: "Women in Budgeting: A Critical Assessment of Empowering Effects, Limits and Challenges of Participatory Budgeting Experiences", Paper for the Asian Regional Conference on Gender Responsive Budgeting Narratives: Transforming Institutions, Empowering Communities, Penang, Malaysia, 24–25 February.

Budlender, Debbie; Sharp, Rhonda; Allen, Kerrie, 1998: *How to Do a Gender-Sensitive Budget Analysis: Contemporary Research and Practice* (Canberra–London: AusAID Commonwealth Secretariat).

Elson, Diane, 1996: "Gender-Neutral, Gender-Blind, or Gender-Sensitive Budgets?: Changing the Conceptual Framework to Include Women's Empowerment and the Economy of Care", Paper for the Commonwealth Ministers Responsible for Women's Affairs, Fifth Meeting, Port of Spain, Trinidad, 25–28 November.

Frey, Regina, 2008: "Paradoxes of Gender Budgeting", Discussion Paper 14. First International Conference on Gender Responsive Budgeting and Social Justice.

Frey, Regina, 2014: "Interlinking Gender Responsiveness and Participation in Public Budgeting Processes: Limits and Opportunities", Paper for the Asian Regional Conference on Gender Responsive Budgeting Narratives: Transforming Institutions, Empowering Communities, Penang, Malaysia, 24–25 February.

Grown, Caren, 2005: "What Gender Equality Advocates Should Know about Taxation", *Spotlight*, Draft (Toronto, Mexico and Cape Town: Association for Women's Rights in Development, October).

Hattam, Jennifer, 2012: Economic Crisis Hits Gender Budgeting, *Inter Press Service*, 22 April; at: http://ipsnews.net/2012/04/economic-crisis-hits-gender-budgeting/ (25 March 2015).

Illo, Jeanne, 2010: *Accounting for Gender Results: A Review of the Philippines GAD Budget Policy* (Quezon City, Philippines: Miriam College).

Jhamb, Bhumika; Mishra, Yamini; Sinha, Navanita, 2013: "The Paradox of Gender Responsive Budgeting", in: *Economic and Political Weekly*, XLVIII,20: 35–38.

Klatzer, Elisabeth, 2008: "The Integration of Gender Budgeting in Performance-Based Budgeting", Paper for the Public Budgeting Responsible to Gender Equality Congress, Bilbao, Spain, 9–10 June.

Klatzer, Elisabeth; Stiegler, Barbara, 2011: *Gender Budgeting—An Equality Policy Strategy*. Briefing paper No. 12 (Shanghai: Friedrich Ebert Stiftung).

Lochhead, James, 2014: *Conference Report Asian Regional Conference on Gender Responsive Budgeting Narratives: Transforming Institutions, Empowering Communities* (Penang, Malaysia: PWDC).

Ministry of Women, Family and Community Development and United Nations Development Programme, 2005: *Manual on Gender Budgeting in Malaysia* (Kuala Lumpur: Ministry of Women, Family and Community Development).

Mishra, Yamini, 2011: "Breaking New Frontiers for Gender Responsive Budgeting: The Kerala Model", August (UN Women).

Mishra, Yamini; Sinha, Navanita, 2012: "Gender Budgeting in India: What Has Gone Wrong?", in: *Economic and Political Weekly*, XLVII,17: 50–57.

Ng, Cecilia; Maznah, Mohamad; Hui, Tan Beng, 2006: "An Unholy Alliance? Women Engaging with the State", in *Feminism and the Women's Movement in Malaysia* (Abingdon, Oxon: Routledge): 63–83.

Phillips, Anne, 1995: *The Politics of Presence: The Political Representation of Gender, Ethnicity and Race* (Oxford: Oxford University Press).

Quinn, Sheila, 2009: *Gender Budgeting: Practical Implementation*. Handbook (Strasbourg: Council of Europe, Directorate General of Human Rights and Legal Affairs).

Rai, Shirin, 2000: *International Perspectives on Gender and Democratisation*. Women's Studies at York (London: Macmillan Press).

Satriyo, Hana, 2014: "Strengthening Vertical Accountability for Gender Responsive Budgeting", Presentation for the UN Women Asia Pacific Consultation on Making Budgets Accountable to Women, Bali, Indonesia, 30 September–1 October.

Sharp, Rhonda, 2003: *Budgeting for Equity: Gender Budget Initiatives within a Framework of Performance Oriented Budgeting* (New York: United Nations Development Fund for Women [UNIFEM]).

Sharp, Rhonda; Broomhill, Ray, 2013: *A Case Study of Gender Responsive Budgeting in Australia* (London: The Commonwealth Secretariat).

Sintomer, Yves; Herzberg, Carsten; Rocke, Anja, 2008: "Participatory Budgeting in Europe: Potentials and Challenges", in: *International Journal of Urban and Regional Research*, 32,1: 164–178.

Sintomer, Yves; Herzberg, Carsten; Allegretti, Giovanni, 2013a: *Participatory Budgeting Worldwide*—Updated Version (Bonn: Global Civic Engagement, Service for Development Initiatives and Service Agency Communities in One World).

Sintomer, Yves; Traub-Merz, Rudolf; Zhang, Junhua, 2013b: *Participatory Budgeting in Asia and Europe: Key Challenges of Participation* (Houndmills: Palgrave Macmillan).

Tan, Pek Leng, 2011: "Beyond Numbers: Women and Politics in Malaysia", in: Ng, Cecilia; Noraida Endut; Rashidah Shuib (Eds.): *Our Lived Realities: Reading Gender in Malaysia* (Penang: Universiti Sains Malaysia): 80–106.

United Nations Development Fund for Women (UNIFEM), 2000: *Progress of the World's Women*, UNIFEM Biennial Report (New York: UNIFEM).

Part I
Gender Responsive and Participatory Budgeting: Narratives of Change and Innovation

Participants holding discussions at a workshop session. The photo was provided by PWDC who granted permission to include it in this volume

Yumiko Yamamoto (Programme Specialist, UNDP Asia-Pacific Regional Centre) posing a question to the speakers. The photo was provided by PWDC who granted permission to include it in this volume

Chapter 2
Interlinking Gender Responsiveness and Participation in Public Budgeting Processes

Regina Frey

Abstract This chapter examines the links between participatory budgeting and gender budgeting discourses, analysing similarities, differences and tensions. What can actors involved in participatory processes learn from a gender discourse, and what can actors working to achieve gender equality learn from discourses on participation? Assuming the main objectives of gender responsive budgeting are gender equality, greater transparency, empowerment of disempowered social groups and more effective budgeting, this chapter discusses the challenges and opportunities for social change in these processes in light of the German experience.

Keywords GRB · Germany · PB · Gender equality · Participation · Budget cycle

2.1 Introduction

Participatory budgeting (PB) is not automatically gender just; and gender responsive budgeting is not automatically participatory. Participatory approaches and methodologies have been criticized for being gender naive (Guijt/Shah 1998: 2), while gender budgeting (GB) could be criticized for a technical (if not technocratic) approach driven by expert knowledge that does not take into account the needs and interests of the most marginalized groups. This chapter argues that both critiques are valid but have shortcomings.

Two examples from Germany are used to highlight situations where GB and PB have both been implemented. The first is Berlin, one of 16 federal states in Germany, and the first government to start GB. It was implemented in Berlin ministries and at the community level (in the 12 Berlin boroughs) and had participatory elements from the beginning. The borough of Lichtenberg is of special interest with respect to GB and PB, as it was one of the first administrative bodies piloting the former in the early 2000s. It was also a pilot community for PB.

Regina Frey, Gender Expert, Genderbüro, Berlin, Germany. Email: frey@gender.de.

C. Ng (ed.), *Gender Responsive and Participatory Budgeting*,
SpringerBriefs in Environment, Security, Development and Peace 22,
DOI 10.1007/978-3-319-24496-9_2

However, the two processes of GB and PB were not interlinked (Schubert-Lehnhardt/Viola 2006). Another interesting case in terms of the intersection of gender and PB is the south German city of Freiburg, where PB and GB were implemented in the late 2000s.

The next section examines PB through a gender lens. The third section analyses GB from a participatory lens, including discussing the different possibilities and entry points for participation in the budget cycle. The conclusion summarizes the arguments in the chapter in terms of the opportunities and challenges in the context of the German experience.

2.2 Participatory Budgeting Through a Gender Lens

In their book *The Myth of Community: Gender Issues in Participatory Development*, Guijt/Shah (1998) state: "Despite the stated intention of social inclusion, it has become clear that many participatory development initiatives do not deal well with the complexity of community differences including age, economic, religious, caste, ethnic and, in particular, gender" (Guijt/Shah 1998: 1). This observation, made in light of the experiences of Participatory Learning and Action (PLA) towards the end of the 1990s and in the context of development work,[1] can also be applied to current PB processes all over the world. In spite of many different participatory and citizens' budgeting processes, examples of gender blindness or superficial inclusion of gender issues abound. For instance, 'taking a gender perspective' has come to mean only efforts to ensure equal access for women and men or counting the numbers of women and men involved in participatory processes.

If, however, participation should be more than just involvement, i.e. if the aim of introducing a 'people's' budget' is empowerment, then it is important to note that participation per se does not ensure a transformation of power relations. On the contrary, adverse effects may arise where research has shown that "participatory spaces can merely reinforce old hierarchies based on gender, caste or race. They can also contribute to greater competition and conflict across groups who compete for the recognition and resources in new ways" (Development Research Centre on Citizenship, Participation and Accountability 2011: 7). This statement can also be applied to projects in Europe, as observed by Massner who pointed out that in Germany, it is 'middle-aged, well-educated men' who are involved in PB:

> One general point of criticism directed at all processes of citizen participation is that these processes provide individuals and influential interest groups who are in any case particularly politically active, with additional opportunities to get involved and influence things. At the level of individuals, this primarily means men from better-educated, higher-earning strata, who not infrequently dominate the participant groups. At the level of groups, this means well-organized interest groups and 'grassroots elites' that are usually both financially strong, and have human resources or time at their disposal. The criticism that additional

[1] See also Cornwall (2000).

participatory offerings enable these segments of society to gain disproportionate and basically undemocratic influence (…) also applies to numerous PB processes (Massner 2013).

If a normalization of unequal power relations is to be avoided, the critique of 'gender-naivety' (if not gender-blindness) would have to be taken seriously. A stabilization of unjust power relations can be avoided if:

(a) different social groups of women and men have the same opportunities to have political influence in PB and
(b) gender is included as a structural category in the processes of PB.

The following section will focus on these two dimensions of gender in PB.

2.2.1 *Influencing Participatory Budgeting*

Studies on PB in Germany show a frequent overrepresentation of men aged 25–45. In the German city of Trier, only 37 % of the contributors to PB were women. Online participation figures showed lower representations of women (Massner 2013). According to Stiefel, only 15 % of the votes in an online survey in Hamburg in May 2006 came from women. The survey attempted to identify ways to cut public expenditure; consequently, the proposals represented 'male preferences'. Interestingly, but understandably perhaps, proposals for reduction in public expenditure were in the areas of social welfare, family support and culture (Stiefel 2010).

A gender-based evaluation of PB in Freiburg also showed that women were often underrepresented (Färber, undated). An exception in terms of representation of women and men in PB is the borough of Berlin-Lichtenberg, where procedures and methods for GB were piloted in the 2000s. It introduced PB some years later. Although, as mentioned above, the processes were not interlinked, an evaluation of the participation of women and men in 2010 showed that women's involvement at 54.6 % was in all forms/modalities (online surveys, events and quarterly conferences).[2]

Examining possible inequalities in women's involvement in PB, however, is only one dimension of 'gender'. A deeper analysis would involve ascertaining whether different social groups of women and men would articulate different or similar needs and interests in such a process. Figures from Germany provide a mixed picture. In the example of Freiburg, Färber demonstrates many different priorities for women, men and single parents, the vast majority of whom were women (Färber 2009: 44). Analysing figures from Lichtenberg, Middendorf

---

[2]Source: http://gleichstellung-weiter-denken.de/pdf/17_forum2_johannes_middendorf_gb_im_buergerhaushalt.pdf (15 February 2014).

concludes that setting citizens' priorities depends on geographical location rather than on gender (Middendorf 2010: 18).

Ensuring equal access is also about choosing different means and methods of participation. For example, online surveys are common instruments employed in PB to assess priorities and seek opinions. However, access to and use of the internet depends on various social categories like social status, age and sex. Figures from Germany show that older persons, particularly women, are 'non-liners', that is, they hardly or never use the internet. In 2007, the 50-plus age cohort had 27.8 % women using the internet compared to 44.3 % of men users (Kompetenzzentrum 2007). In 2013, the overall gender gap in using the internet in Germany was almost 10 %.[3] Thus, if online surveys are a major, or the only, instrument within PB, women, particularly the elderly, and the poor will be excluded.

It is important to differentiate between 'involving women' or ensuring equal representation of women and men on the one hand and ensuring representation with respect to impact and control on the other. As noted earlier, the involvement of different social groups in political and governance processes does not guarantee diminished influence and impact on the existence of 'grass root elites'.

2.2.2 Gender as an Explicit Topic Within Participatory Budgeting

A second component to making PB gender-aware rather than gender-naive is explicitly addressing gender issues within the process and providing funds to ensure gender equality. Gender equality has been legally mandated in Germany in that Article 3 of the German constitution requires the State to actively provide for gender equality and equal opportunities. The government has to ensure gender equality as a basic democratic standard. This not only means that resources should be distributed in a gender-just way, but that, for example, public expenditure should be distributed among boys and girls such that they have similar opportunities to obtain a degree. Further, public funds should be provided to implement affirmative action and gender equality programmes and projects.

Germany offers few examples of PB processes that address gender issues in a systematic manner. One of these is the Freiburg PB process. A consultant's evaluation report recommended methods for including a gender perspective in PB within the course of a 13-month cycle. It started with anchoring gender aspects in the planning phase, training key actors of the administration, generating gender-sensitive material for informing citizens and systematically producing data disaggregated by sex and other categories (Färber undated: 138).

To summarize, participatory processes aimed at empowerment do not always consider gender aspects systematically. It should go beyond equal representation of

---

[3]Source: http://www.nonliner-atlas.de/ (15 February 2014).

women and men in PB. Actors involved in PB should know about gender issues and should continuously reflect on the impact and implications of unequal gender relations in the whole process.

A gender-sensitive participatory approach has challenges since it is more demanding and complex (participation has to be organized and co-ordinated), requires certain expertise (like knowledge of participatory approaches and facilitation) and therefore is more expensive. However, since the alternative might be a gender-blind or 'gender-naive' process that could even reinforce unequal gender relations, a gender-sensitive participatory approach should be worth the additional effort.

2.3 Gender Budgeting Through a Participatory Lens

Examining the intersection of PB and GB from a different angle raises a question—to what extent have other ideas and approaches inspired GB? A study of documents on GB reveals that they do not normally refer to participation but focus on technical procedures and expert budgetary knowledge. To understand why participation is not a major topic in many GB initiatives, it is important to consider the history and the origins of this strategy.

A key publication on the GB discourse is credited to Budlender et al. (1998) who present seven tools for gender budget analysis. These tools were mainly research techniques about collecting data, creating documents and coming up with activities to lobby for gender equality. It addressed gender equality advocates and civil society organizations (CSOs) working on women's rights and needed expert knowledge of government regulations and procedures.

Elements of participatory approaches can be found in these documents, for example, in certain tools proposed by the authors. One of these is 'sex-disaggregated beneficiary assessments'; that is,

> …a tool which allows the voice of the citizen to be heard. Potential and actual beneficiaries of a government programme are asked, using a variety of techniques, their views as to whether existing forms of public service delivery meet their needs as they perceive them. These responses are analysed in order to assess the extent to which a government's current budget meets the priorities of women and men. In essence, women and men participants in beneficiary studies are being "asked how, if they were the Finance Minister, they would slice the national budgetary pie" [Elson 1997b: 13] (Budlender et al. 1998: 41).

This clearly introduces participation but the application of these methods is still mediated and facilitated by experts. One of the reasons for the lack of direct participation is that GB originated from a 'macro-perspective' focussing on the national budget, whereas PB has its origins in a grassroots perspective and has been applied within the context of communities. This clearly shows that GB is not participatory *per se*. However, some entry points for participatory methods exist within GB as discussed in the next section.

2.3.1 Levels and Forms of Participation in Gender Budgeting

What does participation then mean in the context of GB and what forms of participation can the implementation of GB involve? A broad definition of participation would encompass the involvement of actors outside the government (directly or indirectly) in political decision making. This definition identifies three groups of actors in GB processes, each contributing different forms of knowledge:

1. Gender consultants: These actors contribute 'gender expertise', that is, they create or demonstrate data and provide information on gender aspects in different fields of intervention. This results in more evidence-based decision making. Contracted by the government, these experts are not necessarily independent, but as scientists and professionals they provide an outside perspective.
2. Lobby groups/gender equality advocates: These provide knowledge on the situation of various social groups in society. Often, they act as a watchdog and monitor governments, insisting on democratic standards and transparency. They also represent parts of the electorate, thereby giving them substantial political power.
3. Unorganized local citizens: They are rarely involved directly, if GB is understood as a means of influencing and managing the budget. However, if GB implementation requires sex-disaggregated data for a beneficiary analysis, this could involve participatory data collection, where citizens could contribute.

A conflation of roles occurs sometimes, not only because each consultant and lobbyist/advocate is also a citizen, but also because these lobby groups and gender equality advocates enhance their knowledge over time and become experts on gender issues. The following section elaborates on the roles of these three groups in relation to the entry points of participation, citing examples in the German context.

2.3.2 Entry Points for Participation

GB is closely linked to gender mainstreaming. In Europe, the dominant definition of GB stems from an expert group of the Council of Europe. It defines GB as 'an application of gender mainstreaming in the budgetary process' (Council of Europe 2005).[4] This suggests that GB is a management process integrating gender issues

[4]Gender mainstreaming has its origins in the Gender and Development (GAD) discourse. Within GAD, various gender-analysis frameworks have been created which again offer a variety of methods and tools (March et al. 1999). Many of these tools are applied in a participatory manner.

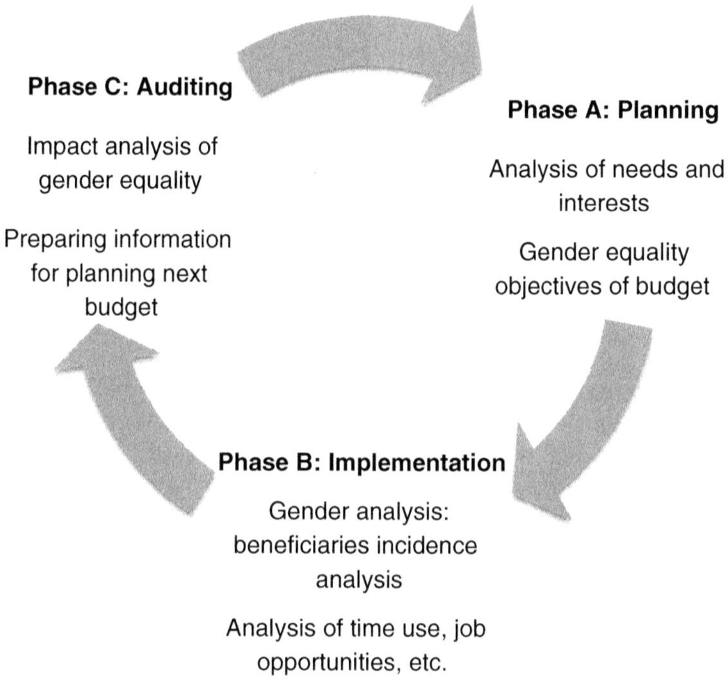

Fig. 2.1 The Budget Cycle. *Source* The author

into the 'mainstream' procedures and regulations of an administration. It is not an additional project and is not limited to one sector of the budget.

As discussed below, the budget cycle is the management cycle for budget planning, implementation and auditing. This section also provides examples of tools that could be used in a participatory manner in the course of this cycle. The budget cycle (Fig. 2.1 above) consists of three phases:

1. Planning: This encompasses budget formulation and its adoption by a government. The result of this phase is a legally binding budget.[5]
2. Implementing: In this phase, the public administration spends money to employ staff, implements different programmes and projects, contracts private companies or hires individuals and so on.
3. Auditing: Monitoring and evaluation takes place during this phase. Was public money spent according to the approved budget (law) and stated objectives?

NGOs/CSOs/gender equality advocates can be involved in steering group committees that monitor GB in all phases of this cycle. For example, in Berlin, a Gender Commission was established at the beginning of the process. Its members

[5]Sharp (2003) differentiates this phase into two stages (preparation and enhancement) and, therefore, identifies four phases within the budget cycle.

comprised high-level administrative staff from various units and from both levels of the Berlin administration (senate and borough). This commission sought advice from gender experts from the very beginning, installing a working group on GB. A member of the NGO (lobby group) 'Initiative for a gender just budgeting in the City of Berlin' was represented in this working group which became the core driver in developing the Berlin GB approach. The members adapted tools for the Berlin context and produced materials documenting approaches and results for a wider public.

In the following section, an example is provided for methods to make each phase of GB more participatory, while highlighting challenges. The example used refers to the public promotion of sports.[6]

Phase A: Planning

When preparing the budget, a government should base its plans on evidence about the requirements, needs and interests of citizens of all genders, age, social status, ethnicity and other categories of social stratification. As mentioned above, one of the instruments in GB is the sex-disaggregated beneficiary assessment that can have participatory elements (Budlender et al. 1998: 41).

The Berlin government published a gender equality framework for 2008–2011, referred to as 'GPR'. This document lists objectives for gender equality policy at the federal state and community levels. When creating the GPR in 2007, the administration organized thematic focus groups and a conference inviting gender equality experts and representatives of NGOs involved in women's programmes and projects. The result was a definition of gender equality objectives in a wide range of thematic fields. It complemented the existing gender equality law and was also used for a re-orientation of ongoing gender mainstreaming and GB implementation by the Berlin government with the help of gender equality master plans. At each level, the administration had to regularly report how it contributed to the objectives set in the GPR. A guideline on GB for the Berlin administration was also provided; it refers to the GPR and the master plan. When setting precise objectives in a certain area of intervention, administration staff had to select objectives for expenditure that had to be in accordance with GPR objectives. They also had to indicate how a certain title or expenditure would contribute to fulfilling GPR objectives.

The GPR defines promotion of sports, especially for elderly women and for single mothers, as part of the gender equality objectives since data from surveys showed that these groups were especially underrepresented in sports services subsidized by the government. Figures on the representation of women and men in the Berlin Sports Association (the main recipient of sports subsidies by the state) showed 36.5 % women's participation (Landessportbund Berlin 2014: 53) as well as 27.3 % women in the executive committee (Landessportbund Berlin 2014: 28).

[6]The example is hypothetical in that it does not exist throughout the whole budget cycle. The figures given in this example, however, are based on evidence.

This is an example of how participatory elements can be included in the implementation of GB in the planning stage when it comes to setting gender equality objectives. It also shows that PB and GB can be viewed as complementary. Sometimes, like in the borough of Lichtenberg, PB and GB processes may run at the same time. Citizens are asked via participatory methods (district assemblies, online or household surveys or other instruments) about their priorities. As outlined above, this should be done in a gender-sensitive way. The results of these processes will be a ranking of priorities that a government should set, although it might not be able to go into detail on how this part of the budget should be spent. If, for example, a result of PB would be 'more subsidies for the promotion of sports', the government will be responsible for the *just* distribution of public funds for the promotion of sports. At this point, gender equality objectives can and should be linked to Berlin's gender equality framework as outlined above. Enhancing participatory mechanisms by including a gender perspective also could mean to communicate gender equality objectives as defined by the GPR. Also, gender analysis results could be introduced to citizens, for example on the website introducing the PB process (Schubert-Lehnhardt 2006: 16). With this background information, citizens would be able to draw a decision on an even broader base of evidence.

Phase B: Implementation

During the implementation phase, the public money spent will invariably have gender impacts. With the help of gender analysis methods such impacts can be assessed: Who benefits? Who will have employment opportunities? Who will have to work without being paid for it? An important (but not the only) instrument of GB in this phase is the sex-disaggregated public expenditure incidence analysis. "This tool can be used to provide an assessment of the distribution of government expenditure of a given programme between men and women and boys and girls" (Budlender et al. 1998: 44f). The questions asked and the instruments selected strongly depend on the output area in a budget and the thematic areas it aims to address.

Returning to the sports example: in Berlin, the 12 boroughs regularly assess the product of 'allocation of sports grounds', conducting a sex-disaggregated expenditure incidence analysis. The results of the assessment in Lichtenberg showed that more than 60 % of this product benefited men (Senatsverwaltung für Wirtschaft et al. 2005: 14). A similar pattern was evident in other boroughs. A survey by the Senate of Internal Affairs and Sports in 2008 showed a slight difference in the number of sports activities undertaken by women and men. Men, on average, were more active, with the overall gender gap being less than 4 % (Senatsverwaltung für Inneres und Sport 2008: 12). However, women and men undertook different *types* of sports. Men and boys preferred sports significantly subsidized by public funding (for example, football), women and girls preferred sports like gymnastics which are often provided by private companies (Senatsverwaltung für Inneres und Sport 2008: 17). Because of these different preferences it can be stated that an imbalance existed between the needs of women and men on the one hand and the subsidies provided by the government on the other.

Phase C: Auditing

An impact analysis is usually conducted in this phase. Technically, this is a comparison of the objectives (determined in phase A) with the output or outcome of implementing the budget (phase B). When this ex-post evaluation shows a deviation from the objectives set, a government will have to think about how goals can be met within the course of the next budget cycle or whether objectives should be adjusted.

The sports example showed an imbalance between the objectives set by the GPR (involving more women, especially certain groups, in sports), the needs of women and men involved in sports (less than 4 % difference) and government subsidies. In at least one borough, an almost 20 % difference was noted in the provision of sports grounds in favour of men. This indicates that the government should reallocate funds according to the different needs of different people.

The results of this analysis can be used by gender equality advocates and in the context of other participatory approaches to 'highlight' and demonstrate the gap between objectives and real outcomes. These facts help CSOs make a government accountable for a gender-just allocation of budget.

2.4 Conclusion

While the importance of linking gender and PB is recognized, several challenges have yet to be overcome. As the chapter argues, integrating gender concerns into PB requires time, energy and money particularly when different social groups (across sex, ethnicity, age, etc.) have to be given a voice in terms of transforming power relations (for example, access to and control over resources). Participation is not merely about numeric representation of women and men. It would be difficult to insert participatory approaches especially at the community level into GB since it was not meant to be participatory in the first place. GB originated from macro (feminist) economics which prioritized interrogating national budgets (Elson 1989, 1991). However, certain participatory methods could form entry points in the budget cycle.

One can thus generally conclude that GB and PB are (still) separately implemented and clear connections between both approaches only exist in some cases. However, opportunities to interlink these two important processes definitely exist. Mainstreaming a gender perspective into PB has the potential to provide more precise results by actively including all social groups of citizens and allowing them to voice their perspectives and concerns. Adopting a gender perspective also means raising questions of inclusion and exclusion in decision-making processes. This is not limited to the representation of women and men of various social groups in PB. It also enables an administration to better reflect its procedures and standards.

However, as noted earlier, combining PB with gender issues poses challenges. Dealing with the complexity of social stratification is difficult. Gender is not just about 'women' or about women and men. It is about women and men in various life situations, according to age, social status, ethnic groups and various other identities. Facilitating this diversity to ensure that power relations are more balanced can make participatory approaches very demanding and expensive.

GB actors can also learn from participatory processes. Although it was not designed as a participatory strategy, GB still has elements of participatory methods. Participatory approaches can anchor GB in concretely interpreting gender equality for women and men. For sustainable governance, objectives should be set on the basis of ground realities, and not from merely speculating about the needs and interests of women and men. Several challenges arise when trying to make GB more participatory:

- In Germany, gender equality is mandated by law. If participatory budgeting exercises were to result in an unjust distribution of funds it could contravene gender equality laws.
- The public budget is a complex matter, especially at the national or federal state level. Good examples exist for combined gender and participatory approaches on budgeting at the community level. However, specialized knowledge is needed to evaluate the impact of certain budget policies on gender relations. Forms of direct participation are likely to have limitations.

From the German experiences, this chapter concludes that GB and PB can and should complement each other given the clear commonality between both strategies. However, challenges, if not limitations, must be borne in mind. At the end of the day, (democratically) elected bodies have to be accountable for allocating the public budget in a transparent and gender-just manner. A government should, therefore, make use of the various opportunities GB and PB offer for good and sustainable governance.

References

Budlender, Debbie; Sharp, Rhonda; Allen, Kerrie, 1998: *How to do a Gender-Sensitive Budgets Analysis. Contemporary Research and Practice* (London: Commonwealth Secretariat).

Council of Europe, 2005: *Gender Budgeting. Final Report of the Group of Specialists on Gender Budgeting (EG-S-GB)* (Strasburg: Council of Europe).

Development Research Centre on Citizenship, Participation and Accountability, 2011: *Blurring the Boundaries. Citizen Action across States and Societies.*

Elson, Diane, 1989: "How is Structural Adjustment Affecting Women?", in: *Journal of SID,* 1: 67–74.

Elson, Diane (Ed.), 1991: *Male Bias in the Development Process* (New York: Manchester).

Färber, Christine, undated: *Evaluation der geschlechterbezogenen Ergebnisse. Kooperations-projekt geschlechtersensibler Beteiligungshaushalt 2009/2010 der Landesstiftung Baden-Württemberg und der Stadt Freiburg* [Evaluation of Gender-Related Results of the Joint Project on Gender Sensitive Participatory Budget 2009/2010 of the Landesstiftung Baden-Württemberg and the City of Freiburg].

Färber, Christine, 2009: *Geschlechtersensibler Beteiligungshaushalt. Ergebnisse und Empfehlungen für die Praxis* [Gender-Sensitive Participatory Budget. Results and Recommendations for Practical Use] (Hamburg: Dashöfer).

Gladwin, Christina (Eds.), 1991: *Structural Adjustment and African Women Farmers* (Gaines-ville: University of Florida Press).

Guijt, Irene; Kaul Shah, Meera (Eds.), 1998: *The Myth of Community: Gender Issues in Participatory Development* (London: Intermediate Technology Publications).

Landdessportbund Berlin, 2014: *Handbuch des Sports in Berlin* [Handbook of Sports in Berlin] (Niedernhausen: Ts.: Schors).

March, Candida; Smyth, Ines; Mukhopadhyay, Maitrayee, 1999: *A Guide to Gender-Analysis Frameworks* (Oxford: Oxfam).

Massner, Kai, 2013: "Participatory Budgeting as its Critics see it"; at: http://www.buergerhaushalt.org/de/node/1879 (15 January 2014).

Middendorf, Johannes, 2010: *Gender Budgeting im Bürgerhaushalt* [Gender in Participatory Budget]. Presentation in September, Berlin; at: http://gleichstellung-weiter-denken.de/pdf/17_forum2_johannes_middendorf_gb_im_buergerhaushalt.pdf (15 January 2014).

Schubert-Lehnhardt, Viola, 2006: *Implementierung von Verfahren des Gender Budgeting in einen Gesamtprozess partizipativer Haushaltsaufstellung am Beispiel des Berliner Stadtbezirkes Lichtenberg - oder: Wie Frauen und Männer aus Berlin-Lichtenberg sich an der Planung ihres Haushalts beteilig(t)en und welche Ergebnisse sie dort bei der Herstellung von Geschlechtergerechtigkeit erreicht haben.* [Implementation of Gender Budgeting in the Overall Process of Participatory Budgeting. A Case Study of Berlin's Lichtenberg District – or: Men and Women from Berlin-Lichtenberg Engaging in Budget Planning and Results Achieved in Gender Equality] Rosa-Luxemburg-Stiftung – Gesellschaftsanalyse und Politische Bildung. Seminarreihe Partizipative Haushaltspolitik (Berlin: Rosa-Luxemburg-Stiftung – Gesellschaftsanalyse und Politische Bildung); at: http://brangsch.de/partizipation/dateien/Expertiseend.pdf (1 March 2015).

Senatsverwaltung für Inneres und Sport Berlin (Eds.), 2008: "Sport in Berlin. Untersuchungen zum Sportverhalten" [Investigations into Sports Behavior].

Senatsverwaltung für Wirtschaft, Arbeit und Frauen, Gender Mainstreaming-Geschäftsstelle, 2005: *Dritter Bericht über Gender Mainstreaming und Gender Budgeting in den Senats- und Bezirksverwaltungen im Land Berlin* [Third Report on Gender Mainstreaming and Gender Budgeting in the Senate and District Administrations in the State of Berlin] (Berlin: Senatsverwaltung für Wirtschaft, Arbeit und Frauen); at: http://www.berlin.de/imperia/md/content/senatsverwaltungen/senwaf/gm/bericht3.pdf?start&ts=1133343074&file=bericht3.pdf (1 March 2015).

Senatsverwaltung für Wirtschaft, Technologie und Frauen Berlin (Eds.), undated: *Das gleichstellungspolitische Rahmenprogramm (GPR) 2008–2011. Strategien für ein geschlechtergerechtes Berlin* [The Gender Equality Framework Programme (GPR) 2008–2011. Strategies for a Gender Equitable Berlin] (Berlin: Senatsverwaltung für Wirtschaft, Arbeit und Frauen).

Sharp, Rhonda, 2003: *Budgeting for Equity: Gender Budget Initiatives within a Framework of Performance Oriented Budgeting* (New York: UNIFEM).

Stiefel, Elisabeth, 2010: "Gender budgeting eröffnet neue Sichtweisen" [Gender Budgeting Opens up New Perspectives]; at: http://www.buergerhaushalt.org/de/article/gender-budgeting-er%C3%B6ffnet-neue-sichtweisen (15 January 2014).

Women in Development Europe, WU Wien, Institut für Institutionelle und Heterodoxe Ökonomie (Eds.), 2009: *Wirtschaft anders denken. Handbuch feministische Wirtschafts-alphabetisierung* [Thinking Economics Differently. Feminist Economics Literacy Manual] (Wien: WU Wien).

Chapter 3
Women in Budgeting: A Critical Assessment of Participatory Budgeting Experiences

Giovanni Allegretti and Roberto Falanga

Abstract Budgeting has for too long been considered a technical arena for highly skilled elites. Participatory Budgeting (PB) opens up the field and creates a space for local communities to discuss the equitable distribution of resources. However, gender has not been at the forefront of the PB debate. On the other hand, gender responsive budgeting has had its own growth trajectory, often not including participatory methods. The chapter highlights possible intersections between PB and gender mainstreaming and notes PB's potential in addressing issues of gender mainstreaming and social justice, following dialogues with other complementary democratic innovations.

Keywords Participatory budgeting · Porto Alegre · Democratic innovation · Co-governance

3.1 Introduction

Participatory budgeting (PB) is today considered one of the most successful democratic innovations of the past 25 years, with almost 2,800 active instances around the planet (Sintomer et al. 2013). Since the first experience took shape in Brazil at the end of the 1980s, PB has been considered as a mechanism to promote

Giovanni Allegretti, Senior Researcher at the Centre of Social Studies, Faculty of Economics, Coimbra University, Portugal. Email: allegretto70@gmail.com.

Roberto Falanga, Postdoctoral Fellow, Mobilizing the Potential of Active Ageing in Europe, European FP7 Project, Institute of Social Sciences, University of Lisbon, Portugal. Email: roberto.falanga@hotmail.it.

C. Ng (ed.), *Gender Responsive and Participatory Budgeting*,
SpringerBriefs in Environment, Security, Development and Peace 22,
DOI 10.1007/978-3-319-24496-9_3

trust and overcome the legitimacy crisis of representative institutions by encouraging citizens' participation and co-governance in decision making on economic and financial issues.[1]

While the variety of PB designs has not been sufficiently studied, PB could likely be best described as an *ideoscape*[2] (Appadurai 1991). This suggests a model that travels around the world and becomes real only through local experiments. Its diversity contributes to continuous change through concrete localized implementations and the different meanings ascribed to PB, according to specific instruments and procedures used to shape its organizational architecture. In spite of these variations, global PB experiments have some minimum common denominators and pivotal principles that make PB recognizable among other participatory innovations that dialogue with it.

Participatory budgets are regarded as important innovations and experiments that investigate new possible forms of governance. Their intrinsic value seems higher in the aftermath of the international crises affecting economies and the legitimacy of representative institutions that appear incapable of challenging and regulating markets. PB could, therefore, become an important space to discuss the distribution of (scarce) resources. It could also contribute to the repolitization of a field—that of budget elaboration—that has for long been considered a mere technical reservoir for highly skilled elites, increasingly gaining strategic importance in public deliberation. PB seems to offer an opportunity to challenge the vision of a neoliberal economy as an inescapable destiny, establishing a richer vision of economic sciences as a field of alternative possible choices that could be addressed in different ways.

Moreover, it is time today for deeper reflections on possible interrelations between PB and its potential in addressing issues of gender mainstreaming and social justice. This issue is undoubtedly relevant, especially if we examine the 25 year history of PB. This paper aims to depict the substantial lack of interest in possible intersections between PB and gender mainstreaming policies. The next section clarifies some issues related to gender sensitive approaches and their substantial absence in the history of

[1]This text owes part of its reflections to the project 'Participatory Budgeting as innovative tool for reinventing local institutions in Portugal and Cape Verde: A critical analysis of performance and transfers' (PTDC/CS-SOC/099134/2008, funded by FEDER—COMPETE and FCT). We want to deeply thank Craig Laird for reading the text with patience and correcting its grammar imperfections.

[2]When Appadurai developed his five dimensions for reading global cultural economy (ethnoscape, technoscape, financescape, mediascape, and ideoscape), he tried to demonstrate that globalization is not merely rooted in the expansion of global capitalism within core–periphery models and does not produce only a homogenized global culture. He sought to demonstrate that modernity circulates through geographic, diasporic, imaginary and local spaces producing several irregularities of globalization (Martínez 2012). Under this perspective, the suffix '-scapes' is used to parallel the variable and often uneven terrain of landscapes to that of uneven global modernization. 'Ideoscapes' can be seen as attempts to capture State power and therefore also consist of counter-ideologies in opposition to modern, dominant political discourses. In this light, we imagine PB as an ideoscape, born in Latin America and hybridized during its circulation around the world.

PB development, notably in its dialogue with other complementary democratic innovations. Following this, the third section proposes some counter stream examples of PB—which constitute a sort of journey around the planet—where a gender sensitive perspective introduced consistent and remarkable innovations. On the basis of these experiences, the concluding section provides some recommendations, taking the shape of a research agenda for improvements in PB and its complementarity with other democratic innovations. It concludes with some policy-oriented suggestions for designing a different future for this democratic innovation.

3.2 Is PB Gender Sensitive?

Unfortunately, few of the nearly 2,800 PB cases show a real sensitivity to gender-related issues, except for some cases in Latin America. The first comparative research on European PB experiences in the past decade (Sintomer/Allegretti 2009, 2014; Sintomer/Ganuza 2011) clearly stated that participatory budgets in the region almost never contribute to changing the social roles of men and women. This was despite the claim that almost everywhere, women appear to be involved in them to a considerable degree, often representing 30–50 % of those involved, with an upward trend when the process becomes more institutionalized. This research noted that in most cases, nothing much was done to facilitate equal participation, even when relevant political female figures attempt (or have attempted) to carry the PB idea forward at the national level, almost constructing their political profile on the basis of the participatory theme. This was true of Segolène Royal (governor of the French region of Poitou Charentes and inventor of the most renowned High School PB) and UK ex-Minister Hazel Blears, who strengthened the PB network in her country, giving it national visibility.

In Africa, Asia and (to a lesser extent) North America and Oceania, the situation does not appear more promising. Certain experiences are exceptions (as in China or Australia) when random selection was applied as a main feature of PB procedures, and gender was used as a variable to select participants for budgetary decisional committees. In these cases, the focus has mainly been on the quantitative aspect of women's presence in the participatory processes. Issues related to power relations in society and the equal valorization of women's voices, their ideas and their decisional and oversight capacities were hardly considered.

The question is: what explains the weak commitment of PB to the adoption of a gender sensitive perspective or the inclusion of gender mainstreaming as a pivotal goal? No single factor explains the negative convergence of so many different cities and political and cultural contexts. However, some possible explanations are often recurrent. These include:

(1) Rarely have transparency and accountability been valued as a real centre of interest for PB. This implies that a careful analysis of public spending is generally not associated with measures of affirmative action that promote social inclusion.

(2) Institutions and procedures responsible for overcoming gender inequality remain fragmented and isolated in many local contexts. They are often in charge of specific policies for specific targets instead of being permitted to insert these goals into a larger spectrum of policies. Furthermore, such gender sensitive structures are often coordinated by parties or officials who are considered marginal to the powerful architecture of governing coalitions.
(3) Often gender budgeting procedures are considered *a posteriori* documents that can contribute more to the understanding of dynamics established to fight gender inequalities than to forge such dynamics by creating participatory arenas to set and fund priorities shaping antidiscrimination and gender empowering policies. As a result, they often act more as sort of gender balance. Such a limited vision diminishes the potential of gender budgeting and reduces spaces to influence the transformation of resource distribution when preliminary budgets are being drafted.
(4) Most gender sensitive actions are viewed as being addressed to women instead of opening new forms of dialogue between women and men. Hence, men often continue to act according to patriarchal/chauvinistic approaches, neither being targeted nor involved directly (as beneficiaries and co-producers) in policies and campaigns oriented towards addressing new visions of relations between women and men.
(5) Self-censorship of women (in social contexts) is often disregarded as an indicator of exclusion, because it is presumed to be voluntary. However, from the perspective of the constraints posed by cultural traditions to the transformation of power relations in society, this phenomenon is undeniably dangerous and recurrent even in participatory processes.
(6) Women's equality has rarely been read in the setting of public participatory policy making innovations from the perspective of the impact of multiple exclusions. Thus, it has not only been separated from disability, single parenthood, age, ethnicity, race or socio-economic segregation, but also from gender orientation concerns.

This complexity of reasons underlines the plurality of agents responsible for PB falling short of its potential in terms of effectiveness. Administrative institutions, i.e. their elected officials and technical staff, are not alone in promoting a diluted PB model that lacks real interest in promoting gender equality and women's empowerment. Civil society also actively dilutes PB as do, to a large extent, universities and other research institutions. As clearly underlined by Santos (2007), several studies in the past decades have often neglected data collection and a gender sensitive perspective when analysing participatory processes and evaluating their effectiveness and efficacy, despite setting out to assess their redistributive effects and their capacity for social inclusion.

In spite of recognizing that many social, political and economic transformations owe a lot to the growing role of women in social life (Avritzer 2007) that contributed to re-democratization in many countries and benefited from them (Alvarez 1991), several studies on PB and other participatory tools of governance innovation have not been analysed for their gender sensitiveness. If this is true, we must recognize—with Santos (2007: 240)—that 'the studies on participatory democracy look blind to the gender differences and women's participation' to the same extent that 'feminist studies on women and/in politics seem everyday more focused on the presence of women in representative institutions, and not in the participatory ones'.

Santos (2007: 242), criticizing the weight literature assigns to quotas and the numerical presence of women in representative institutions, agrees with Araujo's (2002: 150) hypothesis that considerable affirmative action has occurred in the context of neoliberal political reforms. In view of this, Santos suggests that a new research agenda on participatory processes must be based on critical theory with feminist perspective. This must start with Scott's definitions of gender as (1) a constitutive element of social relations, based in differences perceived by sexes (Scott 1988: 42), which imagines gender as a category or a variable of analysis of relations, positions and social relations; and (2) a primary field through which power is articulated (Scott 1988: 43–44), and which sees gender as an attribute of culture.

The lessons that Santos takes from the analysis of so many missing opportunities in valuing the presence of gender sensitive elements in participatory processes are interesting. She stresses the need to take into account a minimal numeric presence of women in every process (reachable through quotas and other affirmative action) and an identity feminist politics that guarantees women's expression and interests. The latter elements represent a widely differentiated evolutionary social category in permanent transformation. The presence of women in participatory processes must translate into a real representation and redistribution of their interests which are clearly plural and complex, thus constituting an open question to be constantly re-analysed.

The need to acquire such a complex perspective requires a preliminary act (Alves/Viana 2008: 45), i.e., abandoning the acceptance—absorbed equally by men as well as by women—of a sort of natural incapacity of women to play a role in the public domain and develop a political intervention. Such an acceptance is often so strong that it succeeds in identifying politics, including participation and collective action, 'as something belonging to [the] male world', to the point that men feel almost 'naturally empowered to deal with politics, exert power, occupy public space…'. Likewise, men show progressively less interest in investing time on small decisions that no longer guarantee solid power.

This is evidenced in a recent comparative analysis of participatory budgets in Spain, Uruguay and the Dominican Republic, where Gutiérrez-Barbarrusa (2012) encountered and described a phenomenon of feminization of PB spaces, especially when resources are shrinking.

3.3 Counterstream Experiences that Link PB to Gender Sensitive Approaches

Fortunately, the above reflections do not represent the entire panorama of global participatory budgets. In the past 25 years, institutions at local and international levels, mainly in Latin America, have tried to promote a different approach to PB and goals of gender mainstreaming. Two main types of gender sensitive (or, at least, women sensitive) PB have occurred when:

(1) Local institutions (often stimulated by supra-local or even transnational networks and organizations) promoted studies on the effects of PB on gender equality or to maximize synergies between existing PB and other processes to promote antidiscrimination visions or affirmative action for gender mainstreaming. Unfortunately, although highly interesting in terms of cultural vision, many of these experiences have only been episodic, producing with some exceptions limited effects on the transformation of PB and its output, and rarely have a permanent effect on political and social culture.
(2) Cities promoted (sometimes to raise institutional awareness or as a result of pressure from social organizations) specific measures to increase opportunities for an active and equal participation of women in PB.

Undoubtedly, the latter has been able to offer a series of creative solutions to the difficulty of realizing gender mainstreaming as a side effect of participatory processes and policies that had no such feature among their initial goals. As the INCLUIR (2007)[3] project proved through its networking activities, it is almost impossible to find evidence of PB resulting in social inclusion (not only for women but also for those marginalized due to ethnicity, disabilities, migration, age or sociocultural status) unless it was an explicit goal and specific tools were employed to achieve this goal. This second group of PB cases is generally limited by two factors. The first is that their strategies seem to concentrate on increasing the numbers of women participants by reducing visible barriers to their presence rather than on problematizing the balance and quality of power relations. The second is that they deal with narrow issues of women's participation, rather than focus on a wider gender-related series of concerns, intertwining them with issues linked to ethnicity, age, education, parenthood status, disabilities and so on.

3.4 Including Women in PB

The Brazilian city of Porto Alegre was among the first to try to monitor and study the presence of women in PB since 1990. During the first 15 years of local PB, the NGO *Cidade—Centro De Assessoria e Estudos Urbanos*[4] monitored several aspects of the evolution of this process. Two books were published in 2003 and 2007, the latter being *Olhar de Mulher. A Fala das Conselheiras do Orçamento*

[3]INCLUIR is the acronym of "El Presupuesto Participativo como instrumento de lucha contra la exclusión social y territorial", coordinated by the City of Venice within the EU-funded URBAL programme; Network nº 9 was developed between 2004 and 2006 and is dedicated to Participatory Budgeting and Local Finances. See more at: http://www.comune.venezia.it (15 March 2014).

[4]See: www.ongcidade.org (15 March 2014).

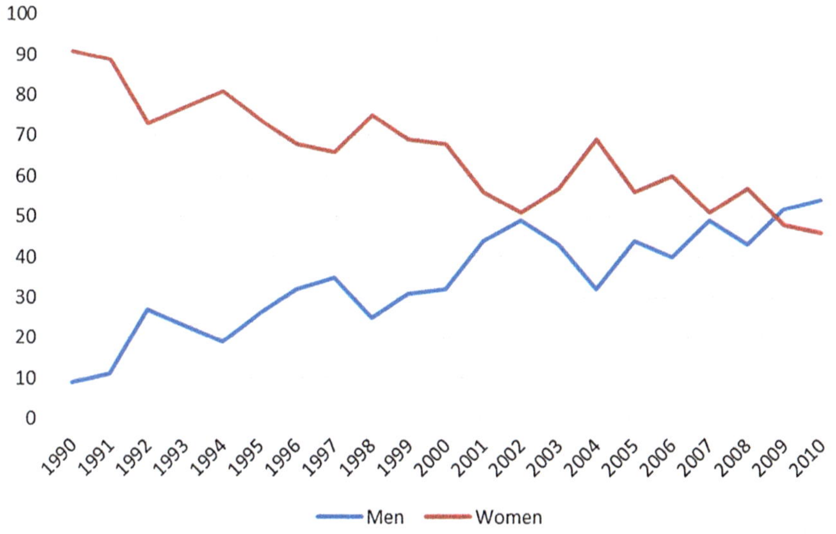

Fig. 3.1 Presence of men and women in the Popular Council (COP) of Participatory Budgeting in Porto Alegre. *Source* CIDADE-PMPA (2010)

Participativo de Porto Alegre[5] (CICADE 2003, 2007). These materials analysed the slow progress of women to participate in the different institutions, with special attention to slum dwellers, that implemented PB. They showed that while women's numbers increased quickly in the basic territorial assemblies, resistance from men who were afraid of losing their positions of power made obstacles to equality more difficult to conquer in the more representative arenas, such as the PB Council or COP which houses the popular councillors elected from the 17 districts.[6] By 2005, women in Porto Alegre already represented 54.5 % of PB participants, but in the COP they only reached such a percentage in 2010 (CIDADE-PMPA 2010).

Interviews with women active in PB revealed several other elements that explained such dynamics and how they were addressed (Fig. 3.1). For example:

(1) Women tended to concentrate on issues linked to family and the quality of social services in areas such as education, health, social assistance and income generation. They accounted for 80 % of the participants in the Health and Welfare thematic assemblies in 2005 (CIDADE-PMPA 2010).
(2) Women seemed to be largely in the 34–60 age group, while men seemed to be largely in the 16–33 and 60-plus age group.

[5]This could be translated as: "Through Women's Eyes. The Speech of Female Councillors on Participatory Budgeting in Porto Alegre".

[6]The COP is made up of 64 citizens elected by the community assemblies during the annual PB process that make the final decisions related to the annual priorities of PB.

(3) Women in leading positions in PB were often single or divorced (62–65 %) and therefore more independent and not requiring men's permissions to participate (Fedozzi 2007).
(4) Many women (especially from the lower social classes) tended to feed prejudices on women's role, often allowing men a monopoly in family representation.[7]
(5) Throughout the first 20 years of PB, women belonging to organized groups (such as NGOs and mothers' clubs) increasingly took part in PB and motivated their members to do the same.

In 2009, such observations were translated into a Manifesto of Action during the Fourth Porto Alegre Conference on Public Policies for Women. Participants envisaged PB as a pivotal political opportunity for building new State-society relations, criticizing the minimal expenditure devoted to specific programmes for women's capacity building.[8] The existence of 170 nursery schools in 2010 which had cooperation agreements with the municipality was presented as a visible indicator of the effectiveness and specific nature of women's struggles within PB. An additional indicator was the creation of several bottom-up enterprises and co-operatives for women and educational committees. However, participants publicly expressed their doubts over the concept of community, emphasized by public institutions, but used to conceal social differences, especially between the sexes. Additionally, they demanded that new programmes be established to empower women to increase their entrepreneurial capacities 'in all the spaces of local and non-local' social life and citizenship, far beyond the traditional emphasis on their role in community organizations.[9] The city of Recife was publicly identified as a model for having created since 2002 a PB thematic Forum for Women that acted as a bridge between policy sectors, linking them to the women's movements rooted in the city.

The transparency and accountability measures guaranteed in all Brazilian participatory budgets including the publication of simplified and understandable versions of general city budgets and multi-annual plans has helped women call for major investments dedicated to their empowerment. This system has been emulated in different cities as a result of pressures from women's movements in different areas of the country and even from abroad.

---

[7]Conclusions presented at the 'IV Conferência Municipal de Políticas Públicas para Mulheres de Porto Alegre'. *Diagnóstico e Desafios*, 11 and 12 September 2009.

[8]Among the data published in the Conference of 2009 that criticized the declining commitment of the Town Hall in promoting women. It was noted that the Porto Alegre women's programme (one of the 21 programmes into which PB is divided) has always been the smallest and most marginal. As an example, it was revealed in 2008 that out of a 2.8 billion budget (in R$), only 109,000 R$ was dedicated to the gender programme, and only 38 % of resources was used.

[9]Conclusion presented at the 'IV Conferência Municipal de Políticas Públicas para Mulheres de Porto Alegre'. *Diagnóstico e Desafios*, 11 and 12 September 2009.

3.5 Mainstreaming Gender in PB

The Brazilian cities of Recife and Fortaleza also conducted specific studies on the presence of women in PB, in order to approve measures to extend gender mainstreaming beyond PB. The central idea of the Recife study was that gender inequalities create different conditions of participation for different subjects. Thus, public policies aimed at gender equality must become structuring elements of PB, whose main value add is breaking the confinement of women to domestic space (Ávila et al. 2006). The subsequent increase in the Women Coordination Commission tasks was considered an opportunity for improving practical features and creating preconditions to increase women's capacity for involvement in PB activities. These programmes included creating specific courses and leaflets on budgeting for women and offering child care facilities during public PB assemblies. The Recife document was also important as it also analysed the limits of a monolithic approach to women's difficulties, stimulating a multi-layered approach to plural and convergent exclusions linked to race and economic conditions of women.

Several of these issues also reappeared in the specific gender sensitive study of PB in Fortaleza (2008) that offered different solutions to similar problems, by creating a Plenary of Segments within PB. This was a special assembly where minority and vulnerable groups such as women who form a majority in numbers but a minority in terms of equality of substantive rights[10] converged. The Fortaleza strategy dealt with specific issues related to women's equality within a wider policy-oriented approach. These were aimed at increasing gender sensitive institutions, creating spaces to support citizens with different sexual orientations and making their representatives meet in the *Encontros da Cidadania* (meeting of citizens), where issues of multiple and multi-layered exclusions were discussed.

The Fortaleza analysis of PB from a gender sensitive perspective was an important step in accumulated knowledge and reflection on the value of struggles which led to the approval of urban equipment and facilities (from kindergartens to health and family care centres) that challenged the sexual division of labour. The end result was a reduction of the overload of work for women (Alves/Viana 2008: 47). However, the study acknowledged the existence of obstacles in convincing male PB delegates of the need to approve proposals strictly linked to their priority and visions and to defend children's interests. Its conclusions therefore promoted strengthening measures capable of reinforcing the dimensions of PB as a space for learning and making women's needs and ambitions more visible, increasing their perceived legitimacy and contributing to consolidate their image as political subjects with full rights by continuously questioning inequalities among sexes and those of different sexual orientations.

These Brazilian experiences became a point of reference in other countries, where gender mainstreaming of PB was not pivotal. In Europe, for example, few

[10]The study *Politicas para as Mulheres em Fortaleza* shows that here (between 2005 and 2008) the women represented 67 % of overall participants.

countries (except the UK) took specific measures to improve women's contributions to PB. The PB Unit was a think tank in the UK that, up until 2012, offered consultancy services to most local PBs. It published a small reflection on the issue (Lavan 2006), referring to the Recife study. It also created special training spaces for immigrant women and provided mobile recreation spaces for children, usually in PB venues in some cities. These steps aimed to facilitate the participation of women who had child care responsibilities.

In general, a deeper inequality between women and men has been recognized and addressed with specific measures in neo-Latin Mediterranean countries. For example, in 2004 in Pieve Emanuele, Italy, a small city forged mainly by migrants from Southern Italy, PB process monitoring underlined the scarcity of women's participation (around 20 %). This was partially because Southern Italian cultures allow men to represent families in public spaces. In an attempt to reverse the situation, the municipality duplicated public meetings, repeating them on the same day at around 5 pm in schools, with a guaranteed extension of children's activities to allow mothers to take part in budget discussions. The strategy achieved excellent results, re-balancing women and men's involvement in PB. In Arezzo and in other Tuscan cities, as well as in Portugal (in Cascais and São Brás de Alportel, for example), mobile play areas and babysitting spaces were organized to allow young families with child care responsibilities to attend PB meetings. In Modena, Italy, an online streaming transmission of public assemblies was undertaken to guarantee the participation of women and young families. In Spain, especially in Andalusia, PBs set specific quotas (of 50 %) for the election of citizens' delegates. Seville, the first city to experiment with quotas for promoting women's representation, inserted PB in a larger political context, and was equally careful to include gender sensitive issues. PB was explicitly linked to the Vice Mayor's Office for Women, but also provided specific support and visibility to LGBT groups and immigrant communities. It must be stressed that Seville, with Fortaleza, is one of the few cities around the world to have dedicated direct attention to gender differences within its specific PB process, recognizing differences between women of different ages.

3.6 Overcoming Cultural Barriers

An interesting experience is from Greater Geraldton in Western Australia. In 2012, the first PB experiment here included a randomly selected committee that guaranteed gender equality. Special meetings were organized for citizens of aboriginal origin, respecting cultural habits, including those related to forbidding direct dialogue among some components of society. During the events, a mobile truck equipped with play areas and computer facilities was offered by the Town Hall to facilitate women's participation.

Among the most interesting experiences in Africa are those of rural villages in Senegal (such as Fissel or Ndiagagnao) where citizens were divided into socially homogeneous groups (women, youngsters, the elderly) to set participants at ease in the discussion of their specific needs and proposals. In places where women in representative democracy did not reach 15 % of the elected officials, PB managed to attract almost 50 % women participation, challenging self-censorship—which usually affects women's participation in large meetings—through temporary separation of the smallest target-oriented groups that could, then, later interact with the overall population.

The *Training Companion for Participatory Budgeting*, published by UN Habitat in 2008 to help disseminate PB in the continent, repeatedly stresses the importance of the cultural dimension, especially gender-biased cultural norms and traditions that influence women's participation in the budgeting process. The handbook calls on local governments to take bold measures to implement women's empowerment and overcome the lack of understanding around local government systems and council management that often affects women more than men. The report notes that some religions forbid women and men from sitting together or, in some instance, working on certain days of the week. In some cultures, one is not allowed to express dissent or criticize higher authorities in public meetings. In others, age is a serious issue where the young people cannot oppose the views of the elders. Therefore, the socio-economic and socio-cultural dimensions call for the local government to be sensitive to diversity among citizens. In addition, effective participation in the budgetary process could be constrained by the language barrier due [to] the multi-ethnic composition of many African countries which calls for the use of indigenous languages during participatory budgeting meetings (UN-Habitat—MDP-ESA 2008, vol. I). Even stronger is this extract (Box 3.1).

Box 3.1: The Case of Singida District, Tanzania

Local tradition and custom holds sway in Singida District. These are often oppressive to women, restricting married women for example from speaking before men, lest they be regarded as prostitutes in the community. Husbands restrict their wives from participating in social and economic activities, and men seize any income generated by women, which leaves them even more dependent on their husbands. Widows may, however, engage in the community decision-making process as they are perceived to be heads of households like men.

It offers some examples aimed at enhancing the links between economic and social policy outcomes and tracking public expenditure against gender and development policy commitments. It also suggests simple measures such as time-tabling PB meetings and choosing venues that would not necessitate significant travel, particularly at night. Scattered grassroots gender budget initiatives that focus on education, health and agriculture are discussed for countries like Malawi, Mozambique, Namibia, South Africa, Tanzania, Uganda, Zambia and Zimbabwe. Specifically, the most quoted African example is that of the Gender Budget Initiative in Tanzania, which resulted in budget guidelines instructing line ministries to submit gender sensitive budgets. Another frequent example concerns Uganda, namely the District Development Project (DDP), promoted at the beginning of the millennium by the Ministry of Gender, Labour and Social Development and the Ministry of Local Government. According to the new strategy, all sub-counties and districts are in possession of well-outlined planning and budgeting guides that emphasize a bottom-up approach to the soliciting of planning ideas and their prioritization. Gender inclusion in planning and budgeting systems and processes, through fair representation of women in public meetings, is emphasized via recommendations for correcting education and career imbalances that require increased education for girls and a cut in the illiteracy rate, currently at an average of 60 % for women and 38 % for men. It has opened the participation of women in non-traditional areas such as the construction of health units and other facilities (UN-Habitat/MDP-ESA 2008: 55). However, most examples of local institutional commitment on gender mainstreaming stem from Latin American experiences. The four most internationally quoted experiences are Cotacachi Canton Municipality (Ecuador), Rosario (Argentina), Santo André (Brazil) and the Peruvian city of Villa El Salvador (Ortiz 2008).

3.7 Three Success Stories

Although the Cotacachi case has been weakened by drastic changes since 2009,[11] the typical romanticizing inertia of networking exchanges continues to consider it one of the world's most interesting case studies of gender sensitive PB due to the outcomes between 2001–2008. Santa Ana de Cotacachi is a municipality located in the Imbabura Province in Ecuador. It has more than 37,250 inhabitants of which 80 % live in rural areas. It has always been marked by ethnic and cultural diversity, of which around 60 % of the population are indigenous Quechua, 35 % white-mestizo, and 5 % Afro-Ecuadorian (Meyers 2005). Its rural population had traditionally been excluded from development processes with poor access to potable water and sewerage. It had one of the highest child mortality rates in the country. With an annual budget of around US$3 million, the municipality was run, until 1996, by the white-mestizo community, as the indigenous majority tended to be politically subordinate, economically pauperized and socially excluded. Furthermore segregation was especially hitting indigenous women in the rural area (UCLG 2011). The election of the indigenous Mayor Auki Tituaña in 1996, and his subsequent re-election in 2000 and 2004, modified local governmental structures, leading to the creation of Women's Coordinating Committees and a series of Annual Cantonal

[11]The main changes—after elections led to a different political coalition—have been in the area of administrative reorganization of political responsibilities, which marginalized participation in the political strategy of the new mayor. For more details see: http://www.uclg-cisdp.org/es/observatorio/la-inclusi%C3%B3n-de-mujeres-ind%C3%ADgenas-en-un-proceso-de-presupuesto-participativo-local (15 March 2014).

Unity Assemblies. The creation of PB in 2002 stated three main intertwined goals: (1) promote social, ethnic, inter-generational and gender-based participation and organization; (2) bring transparency to municipal budget management; and (3) achieve self-management that values the community's economic contribution. An Oversight Committee made up of community members was created in 2003 ensuring that the implementation of jointly decided measures was socially controlled. In the same year, gender-differentiated and positive discrimination measures shaped specific workshops aimed at creating a collaborative environment where indigenous women could feel at ease and could overcome traditional passive behaviour when confronted in a public scene.

As a result, the participation of indigenous women quickly increased and their community organizing capacity was strengthened, through a careful use of their native languages and pedagogical resources employing colours, local symbols and other daily materials. The ascent and empowerment of the women of Cotacachi in PB led to a series of transformations in municipal management, policies and back office procedures. This was supported by special training sessions for municipal technical teams who received specialized skills in participatory techniques and were reinforced by mostly women members. Other transformations included a new arrangement of the Cantonal Development Plan, the Cantonal Health Plan, the Environmental Management Plan, the Parish Plans, and the Community Plans. In 2003, the *Yes I can* campaign was launched, involving 1,667 people, 65 % of whom were women who were taught to read and write (UCLG 2011). Since then, 10 % of all indigenous women and 20 % of all adult women have been taught to read through this programme with the UN declaring Cotacachi the first illiteracy-free canton in Ecuador.

Since the PB application, over two-thirds of municipal resources have been allocated to rural areas, in radical contrast to the formulas previously applied. Significant improvements have been made in rural electrification with 95 % coverage in the sub-tropical area, while 12 % of the annual budget has been allocated to basic sanitation. In a few years, infant mortality has been reduced to zero, and the promotion of traditional medicine was approved, placing value on ancestral indigenous knowledge and equipping informal workers in the area with these skills. As the OIDP Best Practices distinction[12] in 2006 stated, the Cotacachi experience went far beyond the scope of distributing and controlling public resources; it achieved durable economic, political, social and cultural impacts. It underlined the importance of political will as a precondition for fostering the development of a real participatory culture in local society. It showed how institutional actors could empower people, giving them a genuine space to define their policies and control the implementation of their projects. The degree of social integration and sustainability that PB managed to achieve guaranteed the continuation of several of its features, even after the electoral defeat of Mayor Auki Tituaña in April 2009.

---

[12]The OIDP "Best Practices in Citizen Participation" distinction was created in 2006 by the International Observatory of Participatory Democracy. See: http://www.oidp.net/en/projects/oidp-distinction-best-practice-in-citizen-participation/ (15 March 2014).

In the Argentinean city of Rosario, the results of an interesting hybridization of PB with gender sensitive policies were also very promising and sustainable.[13] Its articulation, networking capacity and considerable investments in international diplomacy made it an international model to be emulated. In this case, PB started in 2002, following a methodology adapted from Porto Alegre (Roeder 2010). In 2003, the municipality decided to develop a gender budgeting strategy, soon supported by the UNIFEM gender budgeting programme. This included several different activities for increasing women's involvement in PB. It also included citizen activities such as training civil servants, both women and men, to be more sensitive to gender issues, public campaigns to combat gender prejudices and better interrelated PB and other actor-centred activities aimed at promoting more gender responsible public policies (Bloj 2014). Gradually, all districts were involved in the experiments and a growing number of projects were adopted, with investments of more than US$ 800,000 per year. Among other measures, a system of quotas was established to promote gender equality in the election of citizen delegates in PB. The main goal of such transformation was to disseminate mental change, i.e., a new way of framing public issues in relation to gender (Sintomer et al. 2013). This could be more sustainable than merely increasing women's involvement in PB, which is an important but not sufficient condition, given that it alone cannot transform relationships between women and men in the participatory arena (UNIFEM/UNV 2009).

The UN Development Fund for Women (UNIFEM), which later became UN Women, an agency dedicated to gender equality and the empowerment of women (see: www.unwomen.org), has played an important role in promoting experiences, particularly in Latin America, that try to merge PB with principles of gender responsive budgeting. Rosario and Recife benefited from this support. Today, several cities and social organizations around the world can take advantage of a specific website called Gender and Budgeting.[14] It was developed with the aim of providing a platform for managing and sharing knowledge on Gender Responsive Budgeting experiences in Latin American and the Caribbean.

Finally, it is worth highlighting the positive achievements of one of the latest generations of PB in relation to the issue of gender mainstreaming—that of the United States. In New York, various experiences have benefited from the support of the New York Women's Foundation and community organizations such as Community Voice Heard and WORTH (Women on the Rise Telling her Story). Since early 2011, special attention has been paid to the intersection between gender, economic status, race and age issues. As a result, and as stated in the Second Annual Research and Evaluation Report on Participatory Budgeting (PBP 2013) in New York City, over 60 % of the more than 13,000 who voted on how to spend almost US$10 million of public money in 2012–2013 were women, most being people of colour, Asian, or Latino and lower- or middle-income earners. They were motivated by immigrants and formerly

[13]See *Box 9: Participatory Budgeting and Gender Mainstreaming: The Rosario Experiment* in Sintomer et al. (2013).

[14]See: http://www.presupuestoygenero.net (15 March 2014).

incarcerated offenders, often dispossessed of political rights in the US, showing the inclusive face of participatory democracy and its capacity to address multiple layers of exclusion. An interesting feature that emerged from monitoring participants in New York's PB was that women were more likely to participate in all the stages of the PB process compared to men, as evidenced by the fact that between 2012–2013 women were 66 % of neighbourhood assembly participants, 60 % of budget delegates and 62 % of voters in the PB process. Furthermore, women did not just attend PB events in large numbers, but were also active in their participation: 92 % declared that they spoke during the different phases of PB community organization and during the small group discussions at neighbourhood assemblies. As stated in the detailed analysis of District 39 results (ibid.: 84), community-based institutions have been critical in building trust and engaging women in civic participation. Therefore, it is possible to say that PB challenged the patriarchal paradigm, bringing about a significant increase in engagement, when compared to 2009 local elections where only 53 % of voters were women.[15]

3.8 Challenges to Gender Sensitive PB

PB has long been considered a gender sensitive tool, or at least closely related to gender budgeting and other approaches, in terms of monitoring public finances and studying the impact of revenue and expenditure policy on women and men to stimulate greater gender equality.[16] However, data from international studies shows the opposite; women's participation was much lower than their men counterparts, especially in the higher steps of PB, which usually include some degree of representation and some power in setting the final agenda for PB decisions.

Such romanticizing of PB is possibly linked to its potential, as a result of which it is considered an important tool for the empowerment of social actors traditionally marginalized from decision making in public policies and projects. PB includes a series of distinguishing features such as more profound methodological and communicative process compared to previous participation formulas. These make it seem richer and more radical in challenging norms and addressing the decreasing perceived legitimacy of political/administrative institutions and the individualistic tendencies of society.[17] The imagined components of PB—transparency, accountability and responsiveness—appear to be suitable tools to rebuild mutual trust

[15]In District 39, 97 % of women spoke during small group discussions, 80 % made specific budget proposals and 33 % volunteered to be budget delegates (p. 84 of the 2013 Report).

[16]See: http://www.partizipation.at/part_budget0.html (15 March 2014).

[17]An interesting tale, circulated by the Director of the PB Project Josh Lerner, tells of a woman he interviewed in Rosario for his Ph.D. thesis who proudly affirmed that she was able to 'divorce' her husband owing to Rosario PB. PB enhanced her social relations, allowing her to feel supported, and she could leave the suffering and isolation she lived in when she felt weaker due to lack of friends and community support.

between politics and the social sphere. These would also facilitate tracking expenditure and its impact on social inclusion, including gender equality and empowerment objectives pursued by important international documents such as the Beijing Declaration and the Millennium Development Goals.[18] It is now clear that without specific and coherent measures to make these goals effective, PB could turn into another unfulfilled promise, at least in relation to gender mainstreaming (Bobbio 1987; INCLUIR 2007). Many PB experiences have achieved women's numeric participation; unfortunately, this could prevent action to make them more responsive in terms of gender mainstreaming.

The numeric presence of women in participatory processes can create an illusion of equality, ignoring the differential of power and sociopolitical recognitions between sexes and other gender-related issues. Furthermore, these could overlook the enormous impact of gender-related exclusions when combined with other exclusions related to race, ethnicity, age, parental status, educational or socio-economic conditions (Ribeiro 2000; Martins Costa 2003).

As the Third Gold Report on Decentralization states, 'many of the most dramatic inequalities are related to housing, living conditions and access to basic services, which have knock-on effects on other inequalities, particularly gender inequalities' (UCLG 2013: 111). In this sense, PB is considered a possible solution, capable of triggering and inciting a virtuous circle that can gradually improve living conditions, enhance citizenship, and create feelings of ownership and belonging to a territory among vulnerable social groups (Cabannes 2014). PB also adds value to invisible urban equipment such as underground sewerage networks and water facilities, making them marketable from a political point of view. This allows approaches on basic needs to take on a more central role in the shaping of public policies.

However, these unequal living conditions can hinder the participation of specific social groups if participatory processes do not incorporate concrete measures to overcome them. For example, an important assessment of PB was conducted in Brazil in 2004 by the Inter-American Development Bank (2004) and the Centre for Urban Development Studies of the Graduate School of Design at Harvard University. It showed how timetables and venues of PB public meetings could act as barriers to equal participation by women. It also showed how scale could influence the costs of attending PB sessions and their inclusiveness, especially given that, at the state level and in the municipalities with large rural zones, the gender dimension of participation is particularly striking, since women are more reluctant to travel too far from their homes (Inter-American Development Bank 2004: 25).[19]

---

[18]Especially, see Article 13 of Beijing Declaration and the MDG 3.

[19]Scale and distance have an impact on the participation of women, which falls off rapidly the farther away from the community public meetings are held (Inter-American Development Bank 2004: 38), since physical and financial cost of participation increase with distance from home and affect representation (particularly of women) in the forum of delegates and COP (Inter-American Development Bank 2004: 34).

Some previously quoted studies are specifically interested in analysing the obstacles to women's participation in PB and have revealed multiple reasons for forced self-exclusion from some or all of its different stages. These proved useful in adopting concrete measures to lower access barriers, such as introducing technological tools to facilitate distance-participation. Nevertheless, it has not yet been proved that such measures genuinely work to overcome participation barriers. A study on the Belo Horizonte electronic PB (PMBH 2012) showed that participation among the 25,378 voters of ePB was gender balanced in every age group (49–50 %). It is, however, inconclusive on the advantage for women's inclusion in such a technological rearrangement of PB. The national study conducted by the OPtar project (2013) in Portugal showed that women represented on average 48.8 % of overall participants in public assemblies but only 44.5 % of participants in online activities. The differences between each targeted PB provided inconclusive results in terms of structural reasons for such a dynamic. This suggests that local contexts and conditions weigh heavily on such results and often reproduce in participatory arenas exclusions/seclusions that are strongly rooted in the elected institutions of representative democracy.[20] Similarly, the project has been unable to prove that measures like babysitting facilities have had concrete effects on women's participation, even if it recognizes that they are important for fulfilling rights.

The last examples reinforce the need for further studies and for new methodologies, such as participatory observation, that could offer explanations for the persistent inequality in numbers and, more importantly, in power. Several PB processes have adopted measures of positive discrimination based on quotas, on specific actor-centric processes targeting women, or on the delivery of special training sessions and support materials to increase women's presence in PB. However, detailed monitoring reports and analyses are still missing. These could provide evidence of the effects of such measures not only on numbers measuring women's presence in processes but the quality of their commitment and the resulting output.

Reports published annually in Porto Alegre or New York are interesting starting points but they must be complemented with monitoring reports. These must link the presence of women to their degree of activism, the type and quality of proposals presented, and to their major or minor capacity for attracting general attention and votes of larger audiences during PB voting phases.

---

[20]For example, in PB in Amadora and Leça da Palmeira district in Matosinhos Municipality, the presence of women in public assemblies (reduced respectively to 38.2 and to 11.9 %) can be explained by the fact that mostly members of elected local district councils participate in these two places, thus reproducing the dynamic of Portuguese representative institutions where the presence of women is scarce.

3.9 Conclusion: Towards a More Holistic Research Agenda

Much is still lacking in terms of analyses of the relations between women's movements and the transformation of institutionalized spaces of participation. This would help understand the capacity of social self-mobilization, which Pedro Ibarra (2007) called participation by irruption to influence and modify the spaces of participation by invitation (ibid.) that are often top-down creations. As Santos advocates (2007), such research must adopt and update a feminist perspective and understand the historical relations between State and society in a specific territory, especially its recent improvements through the connection with gender sensitive participatory processes. Such a change would be even more important at a time when a new form of hybridization has occurred between PB and gender mainstreaming to tackle the root causes of inequalities between men and women, encouraging the development of comprehensive programmes that target both men and women, and seeking to change traditional views (Sintomer et al. 2013).

Unfortunately, today, a wave of PB experiences tend towards a hyper-simplification of proposing and voting procedures for fear of losing participants by asking them to fill in too many forms and provide personal data. Such experiences especially active in Portugal and Germany through the use of internet and SMS voting, seriously compromise the possibility of knowing *who* participates in PB. This would make it impossible in the future to set adequate measures to rebalance participants according to their sex, age, race or socio-economic and cultural status.

As stated in Sintomer et al. (2013: 36), 'strangely enough, although they are characterized by elective affinities, PB has not merged with gender mainstreaming very often', except in some instances in Latin America highlighted in this chapter. This can be attributed to the meagre contributions of local political institutions, universities, research centres, social organizations and international institutions that have scarcely supported or stimulated such a merging of models to an innovative culture of gender mainstreaming through PB.

In the past, some important international institutions such as the Urban Management Program of the United Nations, the Friedrich Ebert Foundation[21] and UN Women were active in supporting pilot initiatives. However, in some cases,[22] their regard for gender inequalities was too simplified. Almost 25 years after the first PB experience, the goal of bridging inequalities between women and men still

---

[21]In June 1999, the Friedrich Ebert Foundation in Porto Alegre conducted the first workshop on 'Public Budget and Gender Policies' to strengthen Labour municipal governments to include gender issues in the planning and implementation of municipal policies.

[22]See Indicator no 7 in the UMP document (2004), "Participatory Budgeting: Conceptual Framework and Analysis of its Contribution to Urban Governance and the Millennium Development Goals". It is entitled 'Percentage of women councillors in local authorities' and somehow reduces the understanding of power differences that separate men from women in participatory processes.

seems to lack adequate attention in the transformation and spread of PB around the planet.

One of the most complete documents on gender and PB has been produced by the highly qualified English think tank The PB Unit with the Manchester Women Network (Lavan 2006). It notes the critical need to concentrate on the different uses of the city by women and men, such as the qualitative aspects of equality and the internal deliberative equality (Santos 2007) of analysed PB processes. Moreover, such elements will have to be linked to an integrated and complex interpretation of social exclusion understood in its multi-layered and articulated dimensions. It will also have to be connected to existing interrelations between the active presence of women in participatory processes and the struggles of women's movements in the same territory. Finally, it will be important to try to measure the effectiveness of women's proposals, presented through PB, to change city models and urban cultures, more than just improving single services and urban spaces.

Until now, several of these goals have not been explicitly posed; others have been hesitantly traced. However, the analysis remains inconclusive, the research methods anecdotal instead of evidence-based and, importantly, comparative studies rare. Research could analyse the relationship between PB and gender sensitive issues beyond single case studies and specific contexts to search for common problems and shared solutions among the thousands of PB experiences that are growing daily around the planet.

References

Alvarez, Sonia, 1991: *Engendering Democracy in Brazil: Women's Movements in Transition Politics* (Princeton: Princeton University Press).

Alves, Maria Elaene Rodrigues; Viana, Raquel, 2008: *Politicas para as Mulheres em Fortaleza. Desafios para a Igualdade* [Policies for Women in Fortaleza. Challenges for Equality] (Fortaleza: Prefeitura Municipal de Fortaleza/Coordenadoria Especial de Politicas Publicas para as Mulheres).

Appadurai, Arjun, 1991: "Global Ethnoscapes: Notes and Queries for a Transnational Anthropology", in: Fox, Richard (Ed.): *Recapturing Anthropology: Working in the Present* (Santa Fe NM: School of American Research Press): 191–210.

Ávila, Maria B.; Santos, Joana; Ferreira, Veronica, 2006: *Mulheres no Orçamento Participativo: Estudo da Experiência de Recife, na Gestão 2001–2004* [Women in Participatory Budgeting: Study of the Experience in Recife, 2001–2004] (Prefeitura de Recife: UNIFEM).

Avritzer, Leonardo, 2007: *Relatorio de Pesquisa "Participação e Distribuição nas Políticas Públicas do Nordeste* [Research Report "Participation and Distribution in Public Policies of the Northeast]" (Belo Horizonte: UFMG/Prodep): 9–44.

Bloj, Cristina, 2014: "Participatory Budgets in Argentina: Evaluation of a Process in Expansion", in: Dias, Nelson (Ed.): *Hope for Democracy: 25 Years of Participatory Budgeting Worldwide* (São brás de Alportel: In-Loco Ed.): 135–152; at: https://democracyspotdotnet.files.wordpress.com/2014/06/op25anos-en-20maio20141.pdf (15 March 2014).

Bobbio, Norberto, 1987: *The Future of Democracy. A Defence of the Rules of the Game* (Minneapolis: University of Minnesota).

Cabannes, Yves, 2014: *Contribution of Participatory Budgeting to the Provision and Management of Basic Services at Municipal Level. Municipal Practices and Evidence from the Field.* Background paper for UCLG Gold III, first draft (London: International Institute for Environment and Development IIED).
CIDADE, 2003: *Fazendo Política. Perfil das Conselheiras e Conselheiros do Orçamento Participativo 2002/2003* [Policy Making. Profile of Women and Men Councillors of Participatory Budget 2002/2003] (Porto Alegre: CIDADE).
CIDADE, 2007: *Olhar de Mulher. A Fala das Conselheiras do Orçamento Participativo de Porto Alegre* [Through Women's Eyes. The Speech of Female Councillors of the Participatory Budget in Porto Alegre] (Porto Alegre: CIDADE).
CIDADE-PMPA, 2010: "*Mulheres no Orçamento Participativo: Conquistas e Desafios* [Women in the Participatory Budget: Achievements and Challenges]", in: Atas da IV Conferencia de Politicas Publicas de Porto Alegre para Mulheres, 11–12 September 2009 (Porto Alegre: Preffeitura Municipal de Porto Alegre).
Fedozzi, Luciano, 2007: *Observando o Orçamento Participativo de Porto Alegre* [Observing Participatory Budgeting at Porto Alegre] (Porto Alegre: Tomo Ed.).
Gutiérrez-Barbarrusa, Virginia (2012), "Análisis Sobre la Participación en los Presupuestos Participativos [Analysis of Participation in Participatory Budgets]" in: Allegretti, G. (Compilador), *Estudio Comparado de los Presupuestos Participativos en República Dominicana, España y Uruguay* [Comparative Study of Participatory Budgeting in Dominican Republic, Spain and Uruguay] Malaga: CEDMA, pp. 147–193.
INCLUIR, 2007: *El Presupuesto Participativo como Instrumento de Lucha contra la Exclusión Social y Territorial* [Participatory Budgeting as a Tool in the Fight against Social and Territorial Exclusion], PROYECTO INCLUIR/URBAL 9 and Comune di Venezia, Venezia (Venice: Comune di Venezia).
Inter-American Development Bank, 2004: *Assessment of Participatory Budgeting in Brazil*, (Cambridge, Mass.: Harvard University, Center for Urban Development Studies/Graduate School of Design).
Lavan, Kezia, 2006: *Towards Gender-Sensitive Participatory Budgeting*, Discussion Paper (Manchester: Participatory Budgeting Unit).
Martins Costa, Delaine, 2003: "*A Gramática do Orçamento a partir das Perspectivas de Gênero e Raça: Considerações Preliminares* [The Grammar of the Budget from the Perspectives of Gender and Race: Preliminary Considerations]", Paper for the VIII Congreso Internacional del CLAD sobre la Reforma del Estado y de la Administración Pública, Panamá, 28–31 October.
Meyers, Rodia, 2005: *Cotacachi: Manual de Presupuesto Participativo* [Cotacachi: Manual of Participatory Budgeting] (Cotacachi: Municipio de Cotacachi).
Ortiz, Barbara, 2008: "*Mujer Ciudadana = Mujer Empoderada* [Woman Citizen = Woman Empowered]" (Master Thesis, Lima: PRGEPP/FLACSO).
PBP, 2013: *Second Annual Research and Evaluation Report on Participatory Budgeting* (New York: The Participatory Budgeting Project).
PMBH-Prefeitura Municipal de Belo Horizonte, 2012: *Relatório do Resultado da Votação do OP Digital 2011* [Report on the Voting Outcomes of Digital Participatory Budgeting 2011] (Belo Horizonte: PMBH).
Projeto OPtar, 2013: *Relatorio de Pesquisa do Projeto Optar* [Research Report of the OPtar Project] (Coimbra: CES).
Ribeiro, Matilde, 2000: "*Orçamento Participativo: Panorama Geral e Referenciais Sobre Gênero e Raça* [Participatory Budget: Overview and Benchmarks on Gender and Race]", in: *Proposta*, nº 84/85 (March/August).
Roeder, Eva, 2010: *Der Bürgerhaushalt von Rosario* [Participative Budgeting of Rosario] (Münster: Lit Verlag).
Santos, Cecilia McDowell, 2007: "*Democracia Participativa e Género. Notas para uma Agenda de Pesquisa Feminist* [Participatory Democracy and Gender. Notes for a Feminist Research Agenda]", in: Avritzer, Leonardo. (Ed.): *A Participação Social no Nordeste* [Social Participation in the Northeast] (Belo Horizonte: UFMG Editora): 237–254.

Scott, Joan, 1988: *Gender and the Politics of History* (New York: Columbia University Press).
Sintomer, Yves; Allegretti, Giovanni, 2009: *I Bilanci Partecipativi in Europa. Nuove Esperienze Democratiche nel Vecchio Continente* [Participatory Budgeting in Europe. New Democratic Experiences in the Old Continent] (Rome: Ediesse).
Sintomer, Yves; Allegretti, Giovanni, 2014: *Os Orçamentos Participativos na Europa. Entre Democracia Participativa e Modernização dos Serviços Públicos* [Participatory Budgeting in Europe. Between Participatory Democracy and Modernization of Public Services] (Coimbra: Almedina).
Sintomer, Yves; Allegretti, Giovanni; Herzberg, Carsten; Röcke, Anja, 2013: *Participatory Budgeting Worldwide—A Study* (Bonn: Engagement Global/GIZ).
Sintomer, Yves; Ganuza, Ernesto, 2011: *Democracia Participativa y Modernización de los Servicios Públicos. Investigación Sobre las Experiencias de Presupuesto Participativo en Europa* [Participatory Democracy and Modernization of Public Services. Research on Participatory Budgeting Experiences in Europe] (Amsterdam: TNI).
UCLG, 2011: "Cotacachi, Ecuador: The Inclusion of Indigenous Women in a Local Participatory Budgeting Process—Inclusive Cities Observatory, Barcelona"; at: http://www.uclg-cisdp.org/sites/default/files/Cotacachi_2010_en_final.pdf (15 March 2014).
UCLG, 2013: GOLD III - *Basic Services for all in an Urbanizing World. Third Global Report of United Cities and Local Governments on Local Democracy and Decentralization*—Executive Summary (Barcelona: UCLG).
UN-Habitat; MDP-ESA, 2008: *Participatory Budgeting in Africa: A Training Companion with Cases from Eastern and Southern Africa*—Vol. I and II, Kenya and Zimbabwe (Nairobi/Harare: UN-Habitat—MDP-ESA).
UNIFEM/UNV, 2009: *Experiencias de Participación de las Mujeres para Impulsar el Desarrollo Local con Equidad. Sistematización del Proyecto Conjunto UNIFEM-UNV Presupuestos Sensibles a Género: Visibilizando la Contribución Voluntaria de las Mujeres al Desarrollo de Latinoamérica* [Experiences of Women's Participation to Encourage Local Development with Equity. Systematization of the Project UNIFEM-UNV Gender-Responsive Budgeting: Making Visible the Voluntary Contribution of Women to the Development of Latin America] *Cuaderno de trabajo 9* (Quito: UNIFEM).

Chapter 4
Gender Responsive and Participatory Budgeting in Penang: The People-Oriented Model

Shariza Kamarudin

Abstract This chapter presents the Gender Responsive and Participatory Budgeting (GRPB) pilot project in Penang, Malaysia, under the Penang Women's Development Corporation (PWDC) implemented in collaboration with two municipal councils. It focuses on the component related to community-based projects in two low-cost apartment complexes. The chapter examines the various concepts around participation and their links to the different notions of citizenship; and gender responsive budgeting and participatory budgeting engaging with the community as agents of change. A detailed documentation of the implementation, using the process of dialogical action, then leads to a critical examination of the project's methodology, challenges and innovations, including its impact on women's empowerment.

Keywords Penang · GRPB pilot project · Low-cost apartment complexes · Public expenditure · Citizenship · Dialogical action

4.1 Introduction

Public budgets are meant to be used equitably to ensure social justice for all citizens. People's participation in deciding how public budgets should be spent can enhance the benefits of these resources. It also reflects the government's willingness to apply the principles of good governance in its functioning. If people are to be the beneficiaries of government services, then should they not be part of the process, or even partners, in deciding how public funds are spent?

This chapter examines the specific context of the Gender Responsive and Participatory Budgeting (GRPB)[1] pilot project in Penang, Malaysia. Launched in

Shariza Kamarudin, Senior Project Officer, GRPB Project, Penang Women's Development Corporation, Penang, Malaysia. Email: shariza.kamarudin@pwdc.org.my.

[1]The term GRB was used in the first instance but in the first quarter of 2014, it was changed to GRPB to better reflect the new approach based on the experiences on the ground.

C. Ng (ed.), *Gender Responsive and Participatory Budgeting*,
SpringerBriefs in Environment, Security, Development and Peace 22,
DOI 10.1007/978-3-319-24496-9_4

2012, the project sought to address the needs of its citizenry by amalgamating two models, namely Gender Responsive Budgeting (GRB) and Participatory Budgeting (PB) to influence public expenditure. The framework adopted and the methodology employed dovetail key GRB and PB perspectives. GRB is a responsive process adopted and introduced by policy makers and government agencies as a planning instrument to ensure gender sensitive allocation of resources to promote gender equality. On the other hand, PB is a participatory platform that focuses on empowering communities as stakeholders and agents of change. The vehicle for this was the GRB pilot, a flagship project under the Penang Women's Development Corporation (PWDC) implemented in collaboration with the Penang Island Municipal Council (MPPP) and the Seberang Perai Municipal Council (MPSP), now in its third and final year of implementation.

The chapter focuses on the Penang GRB pilot component related to community-based projects in two low-cost apartment complexes. The PPR[2] Jalan Sungai and PPR Ampangan apartment complexes are owned and managed by MPPP and MPSP, respectively. The first section of the chapter examines the various concepts around participation and their links to the different notions of citizenship. It then examines the concepts of gender responsive budgeting and participatory budgeting in terms of engaging with the community as agents of change. A detailed documentation of the implementation of gender responsive and participatory budgeting in the pilot project is then presented using the process of dialogical action. The concluding section provides a critical examination of the project's methodology, challenges and innovations, including its impact on women's empowerment.

4.2 People's/Citizen Participation

It is important to examine the concepts of participation, people's participation and citizen participation and how they relate to the Penang GRB 'people-oriented model'. The World Bank (1996: xi) defines participation as 'a process through which stakeholders influence and share control over development initiatives, and the decisions and resources which affect them'. From this perspective, participation would be seen in terms of consultation, including participatory decision-making, in all phases of a project cycle: from needs assessment to appraisal, implementation, monitoring and evaluation.

To understand 'people's participation' it is worthwhile to look at Twyman (2000) who argues that interactive participation is a model where people participate in joint analysis leading to action plans and the formation of new local groups or the strengthening of existing ones. Interactive participation involves interdisciplinary

[2]PPR is the acronym for Project Perumahan Rakyat or People's Housing Project. These apartment complexes owned and managed by the Municipal Councils were specifically built to provide housing for the poor and low-income groups.

methods that seek multiple perspectives and use systematic and structured learning processes to achieve the intended purpose. Participants have a say in decisions, and a stake in ensuring the satisfactory delivery of outcomes.

For Wilcox (2003), people's commitment and ownership of ideas are crucial for effective participation; people are committed when they want to achieve something, indifferent when they do not. People are most likely to be committed to carrying something through if they have a stake in the idea. An effective tactic is to allow people to say 'we thought of that'. In practice, it means running brainstorming workshops, helping people think through the practicality of ideas and negotiating with others to arrive at a result acceptable to the maximum number of people.

Cunill (1997, cited in Gaventa/Valderrama 1999) refers to citizen participation as the intervention of private citizens with determined social interests in public activities. Citizen participation in this sense involves direct methods for citizens to influence and exercise control in governance in addition to the more traditional forms of indirect representation. This concept necessitates a redefinition of participation, where people are seen not just as beneficiaries but also as citizens involved in policy formulation and decision making in key areas that affect their lives. Similarly, as noted by Lister (1998), the notion of participation can be expanded to the notion of citizenship, linking people in the political, communal and social spheres and creating new opportunities for agency. For Lister (1998: 228), "Citizenship as participation can be seen as representing an expression of human agency in the political arena, broadly defined; citizenship as rights enables people to act as agents." This chapter will reflect the application of these various concepts of participation in relation to the GRB project in Penang.

4.3 Gender Responsive Budgeting and Participatory Budgeting

GRB is a budgetary practice that considers the various roles of and relations between women and men in society: as husbands and wives, fathers and mothers fulfilling various paternal and maternal responsibilities (Elson 2006). It is, therefore, able to rectify some of the biases in public budgets that generally underestimate the role of women in the macro economy. This is especially true of the unpaid economy where women are almost exclusively responsible for caring, maintaining and expanding the labour force and keeping the fabric of society intact.

Women and men perceive and use their environment in different ways. A study conducted in Manchester found that women and men equally listed crime and lack of order as their greatest concerns. However, while men named the quality of local facilities and transport as important issues that needed to be addressed, women selected the local environment and the quality of education as their priorities (Lavan 2005).

By being responsive to the different gender roles and relationships in society, public authorities can undertake developmental projects targeted at ensuring that citizens can play a meaningful role in fulfilling their needs effectively. GRB is perceived as a public budgeting model that will be able to take into consideration the varied concerns of a 'heterogeneous' populace. It examines the differences among people and how these could affect their needs. Gender intersects with race, ethnicity, social class and income levels, highlighting the roles people play in their communities in the context of their different identities. When planning and implementing public budgets, governments can consider the demands made by individuals through such roles and identities.

Ordinary citizens deserve to know and determine how public funds are being utilized. However, in many countries, budgeting policies are formulated with relative exclusivity, leaving most ordinary citizens without a direct, or sometimes even a representative voice to influence decisions. In some countries, even elected representatives have limited influence on decisions regarding the allocation of public resources. These decisions are the sole domain of a bureaucratic group of people. Moreover, in most countries, public budgeting is traditionally considered the exclusive function of the public administration.

It is only recently that PB has been considered desirable in some countries. A common rationale for budgets being formulated in secret is to prevent fluctuation and volatility in financial and industrial markets. The opposing view is that secrecy in budgeting may give rise to wild speculation while greater transparency may actually create more stable markets (Krafchik 2002). However, PB is about how public funds are used after taking into consideration the opinion of each community member. Therefore, people-centred budgets are also about good governance and social stability (Cagatay et al. 2000).

PB originally started in Porto Alegre, Brazil, in the 1980s. Since then, it has been practised in other countries across the world. PB is one of the most exciting innovations in the development of local democracy. It focuses on community empowerment, recognizing people as agents of change and promoting active citizenship (Sintomer et al. 2013). All persons have the right to have a voice in their community. The marginalized—the poor, children and women—in particular should have the opportunity to identify their needs individually or together as a group and participate in decision making in the allocation of public funds and budgets. Such a process would bring diverse people together, support community cohesion and promote dialogue, equal partnership and ownership between citizens and policy makers. It ensures accountability and transparency as a check and balance on public spending as well as directly improving the socio-economic environment of communities in their areas/localities.

PB may have different interpretations in different countries and contexts. In general, it can be understood as a method that allows non-elected citizens to participate in the conception and/or allocation of public finances (Sintomer et al. 2013). It is a process whereby ordinary members of the public may have a say in how public authorities spend public funds so that expenditure reflects the collective priorities of such people (Institute of Civil Engineers 2005). This is because policy making and public planning are often sheltered from broad public scrutiny and debate.

To be sure, GRB and PB have similarities, as both are concerned with just and equitable allocations, focusing on the interests of marginalized groups in society. Both of these models ultimately affect the nature and type of public projects launched to create the best results for the community. While GRB and PB have overlapping concerns, they differ slightly in their impulses. The former is mainly implemented with government agencies and policy makers as key players in gender sensitive transformations. The latter is bottom-up, benefiting people and communities as key players while government agencies provide enabling roles. By utilizing both approaches in Penang, it is hoped that a model could emerge which demonstrates that government and citizens can come together as partners in influencing and deciding State allocations, resources and development activities.

As will be articulated later in this chapter, engaging with the community as citizens in their own right was found to be highly effective in institutionalizing GRB in the two local councils in Penang and in creating public awareness and confidence in a gender responsive government. The community approach and dialogical action of PB which will be discussed later in this chapter is aimed to complement the GRB process.

4.4 Working with the Community Through Dialogical Action

The framework of dialogical action states that human nature is dialogic, and communication has a leading role in our lives. Through a continuous dialogue with others, individuals create and recreate themselves (Freire 1970). According to Freire, dialogue is the democratic choice of educators. Dialogue allows communication and that is how education occurs.

Freire distinguishes between dialogical actions which promote understanding, cultural creation and liberation, and non-dialogic actions which deny dialogue, distort communication and reproduce power. Dialogical action creates a mutual series of actions and reflections, with both sides acting and reflecting on what is said and done by the other, thus enabling changes to occur. One cannot exist without the other, because action without reflection is like action without thinking, and reflection without action will yield no change or transformation. In the context of sustainable change and transformation, people's minds must change first. The opposite of a dialogical action is a situation where one side regards the other as pitchers to be filled with water. The members of society would be the figurative pitchers, while local government officials and urban planning experts would determine what goes into the pitcher, including when and how.

The community-based component of the GRB pilot project discussed here demonstrates how dialogues with focus groups helped community members focus their attention on their situation with facilitators helping them to mediate their shared problems. Focus group discussions (FGDs) were designed to allow community members to express their needs and concerns. At the same time, they were asked to propose solutions to overcome problems or challenges they experienced in common at their place of residence. This was the turning point where individuals in FGDs, as subjects through GRB intervention, began to shift in relation to how they thought and perceived their world. Community members began to look beyond their individual needs and became agents of change and transformation as they realized that they had a role in influencing public decisions, including in the utilization of public budgets. The following sections examine how GRB and PB were rearticulated and merged in the Penang experience.

4.5 Gender Responsive and Participatory Budgeting (GRPB) in Penang: Merging of Two Worlds

Civil society, through non-governmental organizations, found a political conjuncture in Penang when a new coalition of political parties (Pakatan Rakyat, a People's Coalition) took over the Penang state government in 2008. This created space for new innovations in public administration as there was a desire to do away with the previous status quo. The new government of Penang extolled good governance, competency, accountability and transparency, which also extended to the sphere of public expenditure.

The new Penang State Government set its sights on making Penang an International City. To attain this status, the Chief Minister promised a people-oriented government that would enrich everyone through an equitable share in the economic pie, empower people with rights, opportunities and freedom, and enable people to acquire new skills and knowledge. The expressed commitment to being a people-centred government has certainly made state politicians and bureaucrats more amenable to GRB concepts. PB processes were considered a strategy to leverage on stakeholder interests for the advancement of good governance.

These pronouncements resonated strongly with GRB's principles of equality and participation, although they were not expressed in explicit gender terms. In later budget speeches, the State Government referred to supporting the principles of gender equality and providing allocations accordingly. A tangible demonstration of this commitment was the allocation of RM200,000[3] for the GRB project out of the RM1.5 million for the establishment of the PWDC in 2012 (Ng 2012). At the local government level, favourable political conditions were found in both municipal councils in Penang when the GRB initiative was mooted.

The MPSP Transformation Blueprint 2011–2015 addressed two issues central to GRB: the mindset of the Council (its organizational culture) and the relationship between the Council and the residents of Seberang Perai. With respect to the

[3]USD1 is equivalent to RM4.25.

former, the Blueprint called for council staff to adopt a positive and responsive attitude and to serve the needs and aspirations of the people, the service users. Further, the promotion of gender awareness among the council's management and staff through GRB processes helped council and staff understand different needs and how these could be differentiated between women and men, girls and boys. In terms of the relationship between the Council and the residents of Seberang Perai, the Blueprint stressed the need for an arrangement whereby people would not just be bystanders but would actively participate in the transformation process (MPSP 2011). This provided an excellent context for the pilot to help forge a stronger relationship between MPSP and the communities—women and men—it serves.

MPPP did not have a comparable transformation exercise. Nonetheless, the Council's overall organizational goals and more specific departmental goals provided entry points for the introduction of GRB. Providing quality service featured strongly in MPPP's vision, mission and commitments, while reference was made to the need for community participation. These were the foundations upon which the template for GRB could be built to facilitate the transformation of the council's own budgeting process.

One of the outputs (Output 2) of the GRB pilot focused on the implementation of selected community services within the two local governments. The initial plan was to demonstrate the use of GRB tools to enhance service provision and delivery at the community level. Cleanliness and safety were the main concerns of the state government and councils and these issues actually had the highest number of public complaints. Therefore, it was proposed that the GRB pilot project focus on these areas. Moreover, both areas were part of the Penang State Government's 3Cs (Cleanliness, Congestion and Crime) programme. The cleanliness pilot focused on MPSP's 3R (Recycle, Reuse and Reduce) programme under Local Action 21 (LA21).[4] The other pilot, which was on safety, tied in with the Safe City Programme of both local governments under the National Key Results Areas (NKRA)[5] of crime reduction.

As a result, the pilot project focused on two low-cost apartment complexes, owned and managed by MPSP and MPPP. PPR Jalan Sungai comprises two 22-storey blocks housing 529 three-bedroom units. The average rental is RM110 (USD35) a month. PPR Ampangan is a ten-storey block of 250 low-cost three-bedroom units, also with an average rental of RM100 (USD35) a month. Four strategies were planned for Output 2:

[4]Local Action 21 (LA21) is a mandate to local authorities worldwide to move from agenda to action and ensure an accelerated implementation of sustainable development. The Local Action 21 principle is encouraged to create a society which is healthy and alert towards environmental issues aiming at a balanced socio-economic development.

[5]National Key Results Areas (NKRA) fall under the Federal Government Transformation Programme (GTP) that was introduced in 2010 by the Prime Minister of Malaysia and contains six areas: Reducing Crime, Fighting Corruption, Improving Student Performance, Raising Living Standards of Low-Income Households, Improving Rural Basic Infrastructure and Improving Urban Public Transport.

1. Piloting select services based on common priorities of the authorities and the communities.
2. Engaging the community to provide input on actions to be taken and mobilizing them to participate in implementation, and involving the community in the monitoring and impact assessment of programmes and activities.
3. Employing GRB tools such as sex-based user counts, needs analysis and sex-disaggregated beneficiary analysis to inform decision-making and budgeting priorities.
4. Working with local government staff in both research and action for mutual transfer of knowledge and skills, and building linkages with the community.

In line with these strategies, a project methodology entitled, "Different People, Different Needs: Scoping and Planning to Engender Inclusivity and Ownership in Social Housing" was developed to execute the pilot. It covered four phases: a survey, FGDs, voting and project planning. This is where the Penang GRB merged with PB for the first time through the four phases. The next section discusses the four phases of the GRB methodology.

4.6 The Four Phases of GRB Methodology of Working with the Community[6]

The process began in January 2012 with a reconnaissance conducted in nine flats—seven owned by MPPP, one by MPSP and one by the Penang State Government. This was to decide on the location of the GRB pilot project. The reconnaissance was conducted through observations and conversations between the GRB consultant, the GRB director and the residents of the apartment complexes and site management officers. However, the final location selected for the community pilot was decided by MPPP and MPSP, as the project was conducted in collaboration with (and funding by) the two local councils.

The actual community project started in April 2012 and consisted of four phases (Fig. 4.1). The objective was to understand the needs of the residents and to initiate a dialogue with the local government. It was hoped that this would strengthen local democracy within the low-cost apartment complexes by encouraging the residents to participate in budget decisions and allocation of public funds. This process was also about empowering residents to understand the meaning of shared ownership and to assume responsibility for their own environment.

The four phases were:

a. *Phase 1: Survey*
   A basic household demographic survey of residents of the two PPRs was conducted. One member of each apartment unit provided information on the

[6]The Four Phases of this methodology were designed by Wong Hoy Cheong, the external consultant for GRB from January to December 2012.

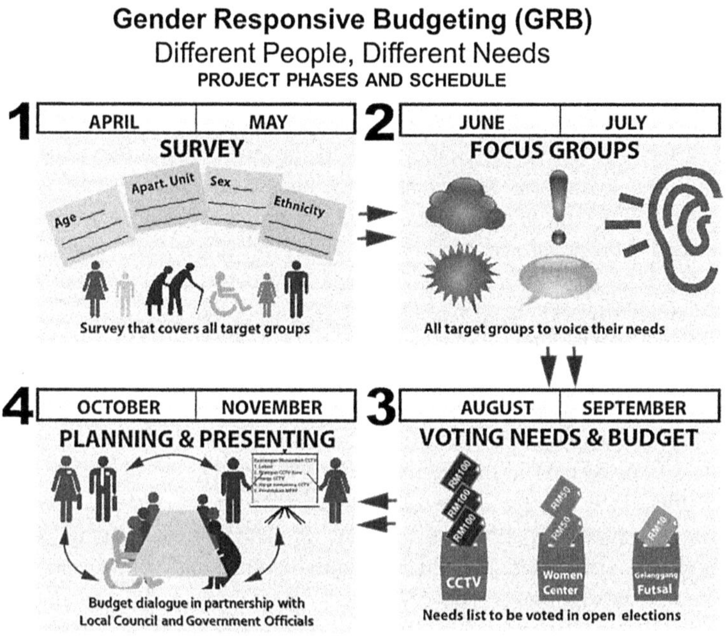

Fig. 4.1 The four phases of GRB methodology. *Source* Penang Women's Development Corporation (2013)

people who stayed with them. The data was keyed in using the Statistical Package for Social Science (SPSS) software programme for analysis and tabulation of the findings.

b. *Phase 2: Focus Group Discussions*
Based on the survey results, FGDs were organized with five groups to understand the needs of both women and men of the major target groups: children/teenagers (aged 18 and below); youths (aged 19–30); adults (aged 31–55); senior citizens (aged 56 and above); and disabled people.

c. *Phase 3: Voting on needs and budget allocation*
This consisted of three days of voting by the residents based on a 'priority needs list' that had emerged from the FGDs. Each resident aged ten years and above was given five ballots of RM100 (USD30) to vote for their prioritized items on the needs list.

d. *Phase 4: Planning, presentation and implementation of project*
Residents sat down with local government representatives to discuss their needs and plan on how to best meet them. Decisions were based on project feasibility, technical considerations and budget availability.

4.7 A Critical Examination of the Four Phases

The project hired four research assistants to assist the GRB team in fieldwork and to work with the communities at the two PPRs throughout the four phases. They were trained as facilitators to identify the needs of the two communities. Their training included role-playing sessions on approaching the residents for the survey and conducting FGDs with the residents. They were also required to write reports on their personal experiences on the ground. This process adopted Freire's dialogical action approach, where all parties—GRB team, research assistants and residents—derived valuable lessons from participating in the journey.

During the implementation of each phase and/or after it was completed, the facilitators and GRB team members reflected on that particular phase of action as illustrated in the diagram below (Fig. 4.2). This process of reflection/evaluation after action and action after evaluation was important to enhance the team's efficiency and effectiveness in conducting the fieldwork. Discussions were held from time to time, face to face, in workshops to prepare for the next phase, or in cyberspace, on matters that needed residents' feedback or on personal experiences.

Community representatives, which included women and young men, especially the residents' associations of the two PPRs, were involved in the implementation of the four phases—assisting in the survey and mobilizing people to attend FGDs and to vote. To highlight the GRB project and to obtain residents' attention and participation in the project, the GRB team together with both local councils—MPPP and MPSP—with the support of both residents associations organized a *Hari Penduduk* (People's Day) at the two PPRs respectively. The event included fun activities, informative exhibition stalls followed by lunch. For example in PPR

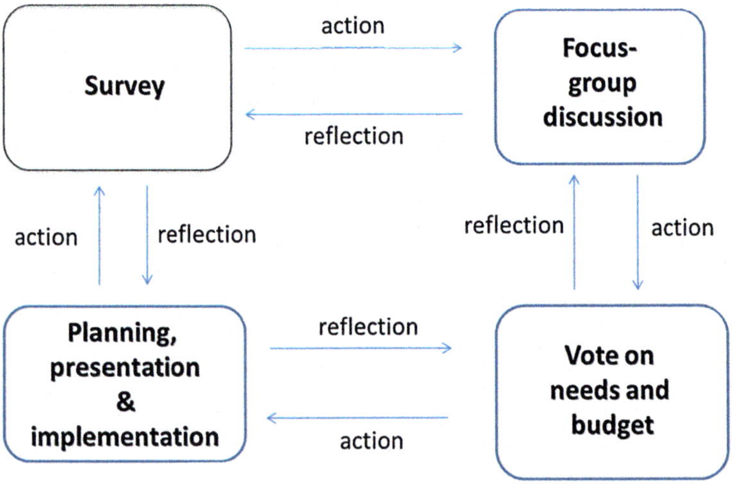

Fig. 4.2 Dialogical action in GRB community pilot. *Source* Penang Women's Development Corporation (2013)

Jalan Sungai, the event consisted of early morning games and a drawing competition for kids; and continued with informative exhibition stalls from the Police, Fire and Rescue Department, Social Welfare Department and Mobile Library for children. At PPR Ampangan, a talk on crime and safety was the highlight of the event. Both events were attended by the Presidents of MPPP and MPSP reflecting support at the highest level of local government.

In the earlier phases, local government employees chiefly attended meetings as representatives of the municipalities. However, they assumed a larger role in Phase 4 by sharing information on the technicalities of the project implemented at both PPRs. This was because Phase 4 related to the implementation of projects in both PPR Jalan Sungai and PPR Ampangan. For the former it was the Community Contract on Cleanliness and for the latter it was the upgrading of the existing Recreational Park. Both projects required many consultations with residents and local council staff to discuss details and ensure compliance with standard operating procedures. The next section discusses in detail the different phases of this methodology.

4.7.1 Phase 1: Survey

The demographic survey provided useful data for planning and designing the subsequent phase. The survey covered 75 % of the units in PPR Sungai Pinang and 85 % of those in PPR Ampangan. Those left out of the survey were residents who were unavailable, despite follow-up visits by the research assistants or members of their residents' association. Some did not cooperate because they were busy, while others were suspicious of the survey since it was conducted prior to the 2013 general election and was perceived as a veiled attempt to study their electoral inclinations.

In both locations, residents aged 18 and below formed the largest group (32 % at PPR Jalan Sungai and 41 % at Ampangan), followed by the 31–55 age group (28 % at PPR Jalan Sungai and 31 % at Ampangan). Those aged 55 and above comprised the smallest group (14 % at PPR Jalan Sungai and 8 % at Ampangan).

Many residents were unemployed, either because they were children, adolescents, retired or jobless at the time. At PPR Jalan Sungai, 33 % fell into this category, while at Ampangan, it was 39.8 %. Housewives made up 12 % of the residents at PPR Jalan Sungai and 4 % at Ampangan. Women, overall, formed the majority of residents in both places, at 53.5 % in PPR Jalan Sungai and 53.4 % in Ampangan.

4.7.2 Phase 2: Focus Group Discussions

The numbers for each FGD were planned (based on the information compiled in Phase 1), to ensure diversity in representation of gender, age and ethnicity in both PPRs. In general, each FGD had to have at least ten people. Based on the information in Phase 1, the FGDs were divided as follows: children/teenagers (aged 18 and below) and adults (aged 31–55) with a total of 40 people in each target group; followed by youths (aged 19–30) and senior citizens (aged 55 and above) with a total of 20 people in each target group. The exception was the group for disabled people which had five individuals. Questions for the FGDs and the answer sheet were jointly designed by the GRB consultant and workshop facilitators/research assistants.

Special care was taken during FGDs with children aged 10–13. Interactive sessions sought to create awareness and educate the children about gender issues in their home and surroundings. Drawing sessions were conducted where the children were asked to draw pictures entitled 'My Family' to identify different members of their families. The children were taken to visit existing sites and facilities within the apartment complexes and asked to classify them as 'Best' and 'Not best'. During the FGDs, questions were posed to the groups by the facilitator and answers were recorded. FGD questions included:

1. What are the issues/problems you experience in PPR?
2. Which groups are affected by these issues/problems?
3. Why do these issues/problems occur?
4. How can these issues/problems be resolved?
5. What is the process for arriving at a solution(s)?
6. Who benefits from resolving these issues/problems?

Facilitators visited residents to extend invitations to join FGDs, ensuring that each floor was represented. However, the FGDs did not turn out as planned. The first few FGDs failed to garner residents' involvement even though they initially agreed to participate in the FGDs. The reason they gave was that they were busy with work or family or had to deal with an emergency.

The experience highlighted the point that it was not easy to mobilize the community without the active participation of its residents. With support from the residents' association, more people were persuaded to attend the rescheduled FGDs. Overall, more women, many of them housewives, aged 31–55 attended FGDs than men and women from other age groups. Snacks were served at most FGDs to encourage participation and to create a more comfortable and conducive atmosphere. In addition, many sessions involving senior citizens were conducted in the living rooms of one of the apartment units. Through the FGDs, a list of the residents' important needs was compiled and classified according to themes/projects/programmes that the community could work on in their own apartment complexes as discussed in the next section.

4.7.3 Phase 3: Voting for Needs and Budget

All residents aged ten and above were eligible to vote. An encouraging 69.5 % of 1,667 PPR Jalan Sungai residents and 67.5 % of 886 Ampangan residents who were eligible participated in the voting process which was accompanied by fun-filled activities such as games, karaoke, dancing and food. These were important in instilling a sense of community and ownership among the residents. A variety of posters and banners was also used to decorate the voting premises to raise awareness about the project among residents. Figure 4.3 depicts the ranking of the voting results according to priority needs and sex.

The ballot papers were marked with a Ringgit (Malaysian dollar) value to remind the residents that they were voting to determine how funds would be spent. Each resident who showed up to vote was given five ballot papers symbolically marked to show a value of RM100 each. The papers were also colour-coded to indicate the age-group and gender of each resident to enable a subsequent analysis of their voting preferences.

At the end of the 3-day voting process, most PPR Jalan Sungai residents chose building maintenance as their top concern, while PPR Ampangan residents voted for a recreational park as their highest priority.

4.7.4 Phase 4: Project Planning, Presentation and Implementation

While Phase 1 to Phase 3 took about eight months to complete, the final phase of implementing community programmes proved to be the most challenging as it stretched to over ten months of numerous consultations among local councils, residents and the GRB team. This was because of the technicality of implementation, which had to follow standard procedures laid down by local councils, as well as internal problems within the communities. For example in PPR Jalan Sungai, the internal problem was the protest by some residents against the existing Residents' Association. The committee members of the association were accused of not being transparent in dealing with their duties, especially in relation to the management of funds. Those who protested demanded a new group to manage the community contract on cleanliness at PPR Jalan Sungai. After several meetings among the residents, MPPP and the GRB team, it was agreed that the members of the new committee should be elected through a democratic process. An election was duly conducted and the winners formed a new Residents' Association called the Pertubuhan Komuniti PPR Jalan Sungai (PKPJS). The new elected committee comprised an equal number of men and women.

In PPR Ampangan, the construction of the recreational park was delayed because many consultations, including one FGD, had to be organized at the community level to discuss detailed features and plans for the recreational park.

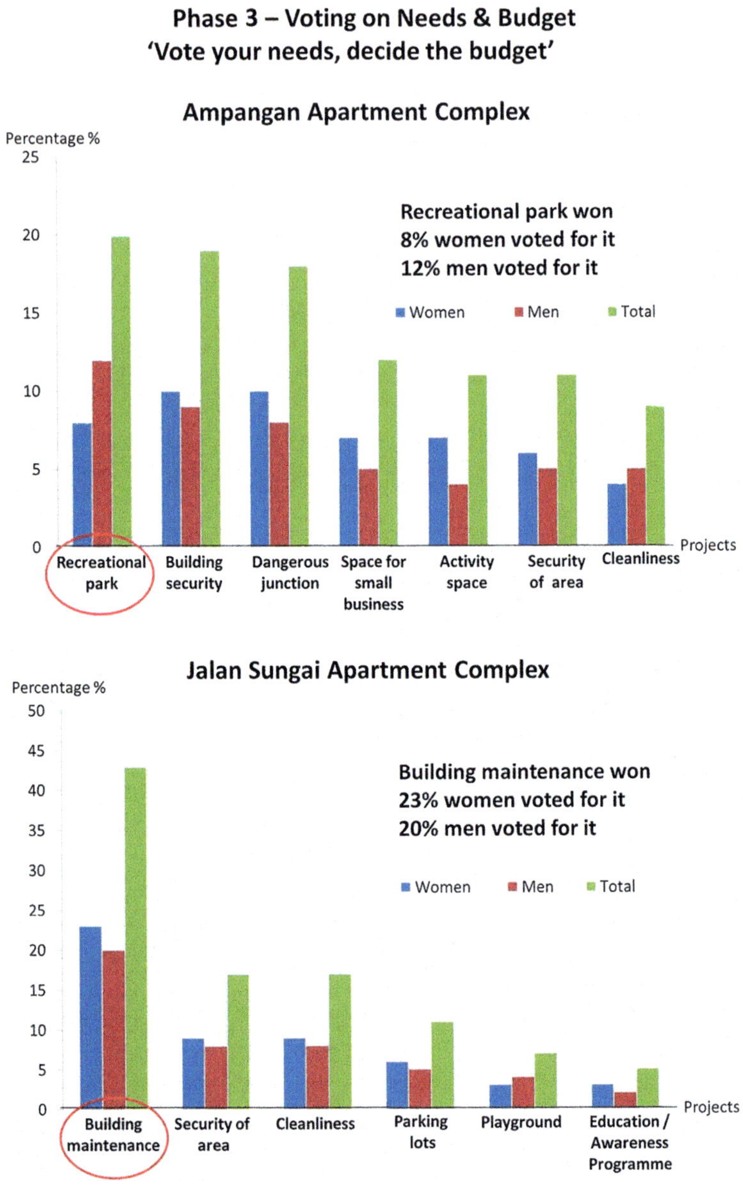

Fig. 4.3 Ranking of voting results according to need and gender. *Source* Penang Women's Development Corporation (2013)

Both consultations and FGDs in PPR Ampangan involved women and men. Another reason for the delay was because many meetings were organized to consult the youth and residents on the futsal court which was one of the facilities requested by the residents, in particular the youth. According to MPSP the futsal court could not be built near their apartment complexes due to limited space and funds. The proposal made by the GRB team to utilize the following year's budget allocation to build the futsal court at the Taman Bandar (City Park), a public recreational park located across their apartment complex, was not accepted by the residents and youth. They were keen to have exclusive use of the facility. They were also keen to enjoy the use of the futsal court as soon as possible. However, after a few more consultation meetings among the residents, MPSP and GRB team a decision was made to drop the plan to build the futsal court.

4.8 Implementing the GRB Project at PPR Jalan Sungai

PPR Jalan Sungai underwent a major cleaning-up and refurbishment exercise during Phase 4 of the project, made possible by a generous special budget allocation from MPPP. Council expenditure for the year for PPR Jalan Sungai rose 300 % over the previous year, totalling RM868,000. The extra allocation of RM700,000 made by MPPP was in response to the needs of residents as expressed through the FGDs, mainly for repainting the two blocks of apartment complexes.

Three MPPP councillors—two men and a woman—who were actively involved and had been supportive of the GRB project from its inception endorsed this idea at the MPPP level and actively lobbied for a budget allocation to be made for the repainting project of PPR Jalan Sungai. Apart from that the other extra allocation for PPR Jalan Sungai was utilized to cover cleaning, refurbishing fire-fighting equipment, repairing and upgrading elevators, refurbishing a dilapidated building used by the community's social organizations, and purchasing new equipment for the children's playground and outdoor gym equipment for adults. According to MPPP the budget for the playground and outdoor gym equipment was derived from the surplus of the 2013 budget.

Continuing with the participatory theme, the residents of PPR Jalan Sungai were asked to vote for the colour schemes they wanted for the external walls of their apartment complexes. Priority was given to those who were regular in rent payment. The objective was to create a feeling of ownership and shared responsibility about their residential area. It was also hoped that this would encourage the residents to pay their rents on time so that they could be involved in other projects in future.

In this apartment complex, the most significant output of the community programme was the awarding of the cleaning contract to the new Residents' Association. The common practice was for such cleaning contracts to be awarded to external private contractors.

As noted earlier, through the voting process, 'building maintenance' had scored the highest in the residents' list of priorities. However, the sum quoted for a complete building maintenance contract was rather large. The MPPP then decided to focus on a contract that was more manageable. It was felt that handling a building maintenance project required experience and management skills which the newly formed committee had yet to acquire. Since the building maintenance also includes the management of the cleanliness of the apartment complexes, MPPP was open to the idea of the Residents' Association taking on the cleanliness contract first as the pilot programme for PPR Jalan Sungai.

After a series of meetings between the local councils and residents, and after the setting up of the newly formed committee, the cleaning contract was awarded to the new residents' association in September 2013. The cleaning supervisor and a team of six cleaners, three women and three men, were hired from among the residents themselves (Fig. 4.4). A panel consisting of representatives of the new residents' association, the MPPP, and the PWDC interviewed the applicants.

The rationale for adopting this community cleaning contract was to leverage on the stakeholder interests of the residents, to provide jobs for the community and to

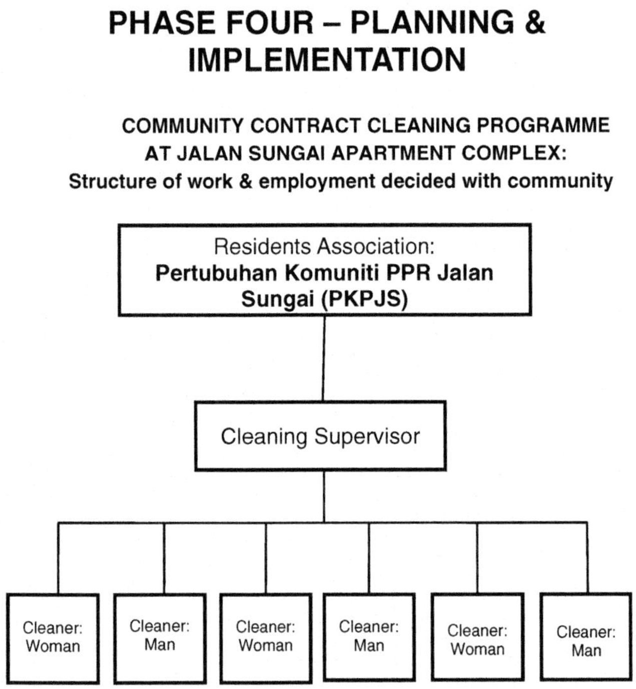

Fig. 4.4 Community contract on cleanliness—structure of work and employment. *Source* Penang Women's Development Corporation (2013)

Fig. 4.5 Impact of Phase 4 of pilot project. *Source* Penang Women's Development Corporation (2013)

instil a sense of ownership and accountability for their residential area. Through subsequent discussions with the Residents' Association, the budget for the contract was increased from about RM66,000 to about RM108,000. The increment was to hire three more cleaners to add to the three hired at PPR Jalan Sungai. The result was great. The cleanliness at PPR Jalan Sungai has transformed very significantly. It is cleaner than before and MPPP was very satisfied with the result (Fig. 4.5). This goes to prove that when residents are given the chance to take part in the process of the planning and budgeting and be responsible for their own surroundings, there is a greater sense of pride and ownership among them.

The cleaners with their innovative and entrepreneurial spirit have gone an extra mile by offering a 'garbage pickup service' for the residents who are willing to pay a small fee of RM5.00 per month. Those who subscribed for this service were given a sticker to display outside their apartment unit (Fig. 4.6). This experiment started with 50 units but later grew to more than 250 units thus enabling the cleaners to earn more income.

Fig. 4.6 Impact of Phase 4 of the pilot project. *Source* Penang Women's Development Corporation (2013)

4.9 Implementing the GRB Project at PPR Ampangan

Due to financial constraints, MPSP could not allocate additional funds for PPR Ampangan. In keeping with the priority need expressed through the voting process, the new addition to upgrade the existing children's playground into a small recreational park, comprising two gazebos, four units of outdoor gym equipment and a unit of foot reflexology path, was constructed for the residents. The details of the new recreational park were decided through a FGD where the majority of the participants were women. The different facilities provided were reflective of the different preferences of the different gender groups. While the reflexology path was popular among the adults and elderly women, the gymnasium equipment was mainly used by the adult men and women and the two gazebos were used by all groups (Fig. 4.7).

Fig. 4.7 Impact of Phase 4 of the pilot project. *Source* Penang Women's Development Corporation (2013)

It was not possible, however, to duplicate the GRB processes at PPR Ampangan. The proposal for the residents to work together on a voluntary basis (*gotong-royong*)[7] to build the recreational park was not well received. The residents felt the concept of *gotong-royong* was not feasible owing to disunity and lack of goodwill among the residents. Further, some residents claimed that they had no time to work on the project and suggested instead that it might be better to advertise the project and put it up for open tender by an external contractor.

At the local council level, councillors and officers were more concerned about ensuring compliance with procedures for open tender from the contractors and meeting safety requirements. Residents at the apartment complexes who were competent contractors and builders were encouraged to apply for the contract through the tender procedure. However, none of them applied and the contract was awarded to an external company. As such, the final phase of GRPB could not leverage on the residents' stakeholder interest for the construction of the recreational park. It was, nevertheless, completed successfully.

Through informal visits and conversations with the residents at the PPR Ampangan about the recreational park, a few women told the GRB team that they were happy that they could exercise. However, they mentioned that they would only use the gymnasium equipment early in the morning and later in the evening while it was dark because they felt shy if someone was watching them doing the

[7]*Gotong-royong* is the Bahasa Malaysia expression for people or a community to voluntarily come together, co-operate and work together on a specific project without payment.

exercise. Later, a Benefit Impact Analysis (BIA) conducted by the GRB team to document the usage of the park revealed that men and boys were the majority users of the recreational park compared to women.

4.10 GRPB: The Best of Both Worlds

Two years after the implementation of the GRB pilot project, particularly after the experiences on the ground, the GRB team decided that it was important to include 'participation' as part of the project name since the PB process played a vital and equal role as GRB. The name of the GRB pilot project was formally changed to Gender Responsive and Participatory Budgeting (GRPB) project after the Asian Regional Conference on GRB: Transforming Institutions, Empowering Communities that was held in Penang on 24–25 February 2014. Significantly to reflect due emphasis and importance of both approaches of budgeting, the GRB team at PWDC is now known as the Gender Responsive and Participatory Budgeting (GRPB) team.

The GRB methodology analyses decision making on public expenditure in the context of implications on gender equality and attempts to bring development and changes that will make budgets more gender equal. The PB model, on the other hand, introduces processes in which citizens can directly influence aspects of budget decision-making. The GRPB People-Oriented Model of Penang as illustrated in the diagram below shows the amalgamation of GRB and PB where the two types of budgeting processes play a complementary role to make sure gender sensitivity and inclusivity are in place to promote gender equality and social justice.

Local governments—MPPP/MPSP—have a duty to ensure that the planning, budgeting and implementation of their projects and programmes integrate the representation of voices of different groups of people in society, without losing sight of the need to mainstream gender perspectives as well. By institutionalising these processes in all their projects and programmes local governments are in fact putting into practice the principles of good governance.

The methodology of getting the people's participation is through participatory budgeting where the community obtains the opportunity to vote for their needs and budget through a democratic process. The Four Phases of the GRPB Methodology gave people the power to play their roles as active citizens and to become partners—not just beneficiaries—together with the local councils especially in formulating policies and making decisions in the key areas that affect their lives. Both meet in this process of dialogical action and reflection *a la* Freire (Fig. 4.8).

The marriage of PB and GRB in Penang was born out of a desire to engage with the community. For the principles of gender responsiveness to be institutionalized and internalized by the public administration, it was necessary to demonstrate that people would embrace this process. The GRB Project in Penang is, therefore, unique because of its inclusion of the people as stakeholders and in terms of citizenship participation. Combining GRB and PB marks a change from usual GRB models and processes.

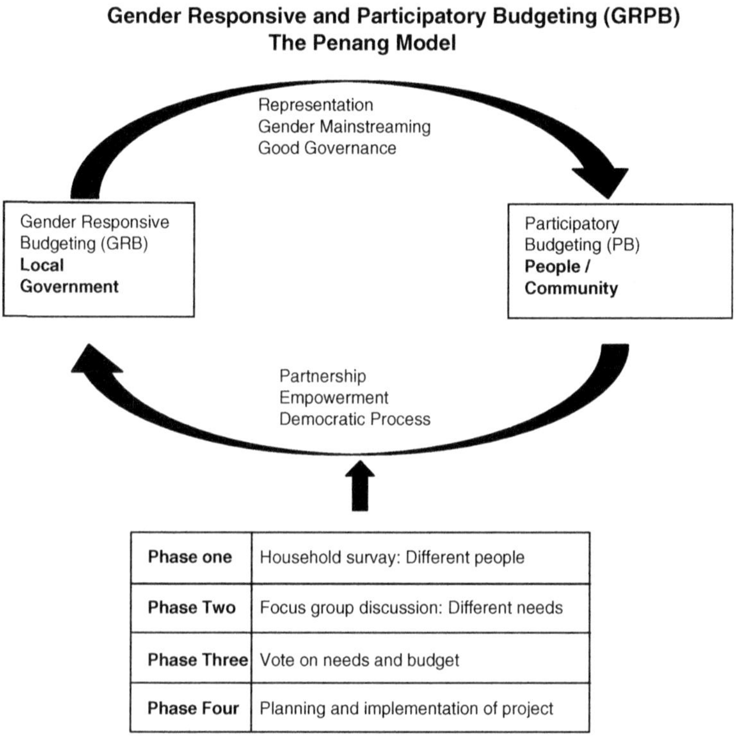

Fig. 4.8 GRPB—The Penang Model. *Source* Penang Women's Development Corporation (2013)

This project was evaluated in November 2013 by Dr. Regina Frey, a German expert on Gender Mainstreaming and Gender Budgeting. The evaluation report indicated that the project's greatest strength was the two community pilots in PPR Jalan Sungai and Ampangan which made the benefits of GRB visible, creating a clear framework at the community level. She concluded:

> With the pilot projects the GRB team developed very useful showcases. A methodology was created on how communities can express priorities in a participatory, democratic and gender sensitive way. It sets four phases and ensures participation of various groups of women and men. The project proves that GRB is useful because the people start to organise themselves and take on more responsibility for their community. Here the GRB project worked with an innovative approach, also serving as a bridge between the communities and the local administration (Frey 2013: 8).

4.11 Women's Empowerment

Women stand a good chance of becoming movers in GRPB projects. In PPR Jalan Sungai and Ampangan, they mobilized others during the Four Phases of the GRPB pilot project and became leaders in many activities and programmes organized at their place. For example, for Phase 1, they mobilized the residents to assist the GRPB team in doing the survey on each floor of the apartment complexes. In Phase 2, they mobilized other women and men from different ethnic groups and age categories to be involved in the FGDs. For Phase 3, the women together with the young men in the community mobilized residents to come down and vote. They also assisted the GRPB team with the registration and voting exercise.

Women who were more confident, experienced and educated took the lead in organising events and were capable enough to be the Master of Ceremonies (MCs) in the various community events. As for those women who had less or little formal education, in particular the elderly women, they played an active role in managing the catering of food and drinks.

Through the experiences of working with the women on the ground, the GRPB team learned of the different gender dynamics at the different PPRs. In PPR Jalan Sungai the women were more vocal and upfront and took important roles whereas in PPR Ampangan the women were quieter and less expressive but nevertheless could perform their duties well and show leadership qualities. In truth, the men in PPR Sungai Pinang were more supportive compared to the more dominant roles played by the men leaders at PPR Ampangan.

The GRPB project at PPR Jalan Sungai has certainly helped open up the spaces for women's empowerment. A new association called Pranita Wawasan was set up by the women in 2013 to focus on skills building, entrepreneurship, awareness and empowerment programmes. The GRPB project also has facilitated the entry of other NGOs to carry out community work at PPR Jalan Sungai, creating a synergy among these groups. For example, the Women's Centre for Change (WCC), a women's group which is based in Penang, conducted a series of awareness programmes on Violence against Women (VAW) for the women at PPR Jalan Sungai. The Soroptimist International Penang (SIP), also a women's group, organizes skill-building classes/sessions on computer skills, cooking and sewing for women there.

Another group that conducts programmes for the children and youth—girls and boys—at PPR Jalan Sungai is Penang Arts-Ed, a community-based non-profit organization that utilizes arts and culture for education and community empowerment. Arts-Ed works with the young girls and boys at PPR Jalan Sungai through activities which focus on skill building and awareness, particularly programmes on videography, dancing, and creating art and craft from recycled items.

4.12 Conclusion

There are several challenges in implementing GRPB at the community level. The most obvious challenge was and still is the inconsistency in commitment and time invested by residents and local council representatives. GRPB processes can be time-consuming because they include many complex layers of planning and consultation between the residents and local councils.

Secondly, negotiating with various local personalities and conflicting, if not fragmented, interest groups in the communities is a long and complex process. Both PPRs consist of diverse and fractured communities that slowed down the Phase 4 implementation process. Often, it was a challenge to get the residents to come together. Some residents disliked certain members of their residents' associations for personal reasons. As noted by Frey, 'These power relations follow (not just) ethnic lines but also social class, gender, age, disabilities and other social categories that sometimes make it difficult to achieve a solid consensus' (Frey 2013: 8).

To bring about some unity of purpose, the GRB team had continuous consultation meetings with the Residents' Associations and residents, always trying to ensure that there was representation from different gender and social groups—in terms of sex, age, and ethnicity. In addition, there were programmes organized to bring together the people at the PPRs. For example, festivals like *Hari Raya, Chap Goh Meh, Hari Merdeka, Nuzul Quran*[8] and events to launch and celebrate the implementation of the Phase 4 projects were attended by the residents from the different gender and social groups. The local council representatives were also invited to these events to mingle with the communities.

A third factor was the lack of expertise including negotiation skills of the residents. Phase 4 included a project for capacity building wherein the end objective was for the residents to present a working paper together with the GRB team and local government at the budget dialogue session. This was to demonstrate a model that could be adopted in the GRPB project where the community presented their needs and priorities that had resulted from the FGDs and voting processes carried out.

However, unfortunately this did not materialize. Apart from the time taken up by the planning and consultation meetings for the implementation of the prioritized projects, the failure can also be attributed to the fact that communities at both PPRs had little or no exposure to this kind of process, where they would have to choose representatives competent enough to present the results of their deliberations at the local council meetings.

They were clearly hesitating and unprepared to do this due to their lack of presentation and negotiation skills. The GRPB team has duly taken note of this matter and has planned a series of capacity-building and leadership workshops for the residents. It is hoped that with this series of workshops the community could gain the necessary skills and confidence to eventually present their needs and priorities at the budget dialogue sessions organized by the local councils.

The local government's approach, while improving, remains less than ideal, another challenge to be considered. While local government officers were deeply involved in planning the projects chosen by the residents and ensured that technical matters were addressed, they still tended to treat the pilot as a project that was outsourced to PWDC. Implementing GRPB requires a change in mind-sets and

[8] *Chap Goh Meh* is celebrated to mark the end of Chinese New Year; *Hari Merdeka,* the national day of Malaysia, is celebrated on 31 August to commemorate independence from British colonial rule in 1957; *Nuzul Quran* marks the 17th day of Ramadan; and *Hari Raya Aidilfitri* marks the end of Ramadan.

procedures need to be institutionalized within the local authorities; this is necessarily a long-term process.

Last but not least, people's expectations also need to be managed. Implementing GRPB with the public can generate a lot of excitement and many expectations. However, the reality of financial limitations still holds true, and when the residents vote for projects that cannot be implemented as a result of such limitations, they may experience disenchantment and frustration. A case in point was PPR Ampangan where a futsal court was not included in the implementation due to budget constraints and lack of space around the apartment complexes area. This frustrated the residents, especially the young men. Therefore, it is vital to manage the expectations of communities, especially in the earlier phases of implementation. To overcome this, a series of consultation meetings were organized among the local councils, the residents and GRPB team with the agreement that the futsal court project would be carried forward to next year.

Notwithstanding the challenges and limitations mentioned above, a number of positive developments have also stemmed from the GRPB project. Local governments are now more willing to make empathetic decisions that benefit the community. When decision makers in the MPPP observed that GRPB had increased community participation and gave voice to people's needs and aspirations, they reciprocated by increasing budget allocations to improve their living conditions. Councillors and officers in both MPPP and MPSP also reacted positively to the residents' ability to organize themselves and express their needs in a democratic fashion. The residents were clearly empowered by GRPB and developed a greater sense of ownership of their shared facilities and resources. Indeed, it reflected the notion of interactive and effective citizen participation as noted earlier by Twyman (2000), Wilcox (2003) and Cunill (1997, cited in Gaventa/Valderrama 1999).

This three-year project culminated in a unique framework that empowered ordinary citizens to participate in public expenditure decisions affecting themselves and their community. They became agents of change and were able to set aside their differences to decide for themselves as a residential community. The experience of accomplishing Output Two—the community projects at PPR Jalan Sungai and Ampangan—has shown that a diverse society can arrive at singular decisions that can be celebrated by the community as a whole. Empowering communities, especially women and the marginalized, for Penang's transformation is crucial towards realizing the vision and mission of the state government.

GRPB allows people to witness and experience the change that they want within their physical locale. This will produce a series of highly beneficial impacts as communities become more committed internalizing a greater degree of civic consciousness. However, for long and lasting change, the way forward is mainstreaming GRPB into local authorities and ultimately institutionalizing GRPB into the public administrative process at all levels of the state. Only then can it be sustainable at both local government and community levels.

Acknowledgments The author would like to thank Dr. Cecilia Ng, Advisor for GRB pilot project PWDC; Wong Hoy Cheong, Consultant for GRB pilot project year 2012; Tan Pek Leng, the Chief Executive Officer of PWDC; Aloyah Bakar and Henry Loh, Director and Manager of PWDC's GRB pilot project PWDC; and James Lochhead, GRB Consultant, for their inputs and technical assistance in writing this chapter.

References

Cagatay, Nilüfer; Keklik, Mümtaz; Lal, Radhika; Lang, James, 2000: "Budgets as if People Mattered: Democratizing Macroeconomic Policies", in: *SEPED Conference Paper Series* (New York: United Nations Development Programme): 2.

Gaventa, John; Valderrama, Camilo, 1999: "Participation, Citizenship and Local Governance", Background note for Workshop on Strengthening Participation in Local Governance, Institute of Development Studies, Sussex, 21–24 June).

Elson, Diane, 2006: *Budgeting for Women's Rights: Monitoring Government Budgets with Compliance for CEDAW* (New York: United Nations Development Fund for Women).

Freire, Paulo, 1970: *Pedagogy of the Oppressed* (New York: Continuum Books).

Frey, Regina, 2013: *Evaluation Report—PWDC's Gender Responsive Budgeting Project in Penang* (Berlin: genderbüro).

Institute of Civil Engineers, 2005: *Citizen Engagement and Public Services: Why Neighbourhoods Matter* (London: Institute of Civil Engineers).

Krafchik, Warren, 2002: "Can Civil Society add Value to Budget Decision-Making?"; in: Judd, Karen (Ed.): *Gender Budget Initiatives. Strategies, Concepts and Experiences* (New York: UNIFEM); at: http://www.unwomen.org/~/media/headquarters/media/publications/unifem/genderbudgetinitiativeseng.pdf (15 March 2014).

Lavan, Kezia, 2005: *Towards Gender-Sensitive Participatory Budgeting* (Manchester: Church Action on Poverty).

Lister, Ruth, 1998: "Citizen in Action: Citizenship and Community Development in a Northern Ireland Context", in: *Community Development Journal*, 33 (3): 226–235.

Majlis Perbandaran Seberang Perai (MPSP), 2011: *Blueprint Program Transformasi Majlis Perbandaran Seberang Perai* (Penang: MPSP).

Ng, Cecilia, 2012: "Right Time, Right Place and Some Luck. PWDC in the Making", in: *Penang Monthly* (April): pp. 18–21.

Penang Women's Development Corporation, 2013: "Report to the GRB Steering Committee. October" (Penang, Malaysia).

Sintomer, Yves; Traub-Merz, Rudolf; Herzberg, Carsten, 2013: "Introduction: Participatory Budgeting—A Global View", in: Traub-Merz, Rudolf; Sintomer, Yves; Zhang, Junhua (Eds.): *Participatory Budgeting in Asia and Europe: Key Challenges of Participation* (London: Palgrave Macmillan).

Twyman, Chasca, 2000: "Participatory Conservation? Community-Based Natural Resources Management in Botswana", in: *The Geographical Journal*, 166 (4): 323–335.

Wilcox, David, 2003: "The Guide to Effective Participation"; at: http://partnerships.org.uk/guide/index.htm (15 March 2014).

World Bank, 1996: *World Bank Participation Sourcebook.* Environment Department Papers Participation Series (Washington, D.C.: World Bank).

Chapter 5
Community Participation and Women's Empowerment: The Kerala Experience of Participatory Planning and Budgeting

Sunny George

Abstract This chapter provides a brief account of Kerala state's decentralization process and its achievements, particularly in relation to participatory budgeting and women's empowerment. The chapter describes the process of development as expanding people's capabilities, and strengthening of local government by devolving functions and finances. With efforts to actively involve people in local governance at every stage, and embed women's empowerment and poverty reduction in the system, the powers and functions of the *grama sabha* are actually embedded in the people. Using case studies of women's empowerment, this chapter highlights the lessons learnt from these reformed structures and processes.

Keywords People's participation · Women's empowerment · Participatory budgeting · Governance · *Panchayat* system

5.1 Introduction

In India, a key strategy for achieving sustainable human development is the development of a good local governance system. Interventions for development, particularly poverty alleviation, are considered more effective when they are initiated through local governments (CDS 1980). The Government of India's commitment to decentralization was explicitly stated with the 73rd (for rural area) and 74th (for urban area) Amendments to the Constitution in 1992 (Government of India 1992a, b). These made local government institutions constitutionally mandatory. State governments were required to take steps to create *panchayats* and municipalities and endow them with powers and authority to enable them to function as institutions of self-government. Thus, *panchayats* in rural areas and municipalities in urban areas came into existence with elected bodies to govern them.

Sunny George, HUDCO Chair Professor and Director, South Asia School of Local Governance, Kerala Institute of Local Administration (KILA), Kerala, India. Email: sunnygeorgekerala@gmail.com.

C. Ng (ed.), *Gender Responsive and Participatory Budgeting*,
SpringerBriefs in Environment, Security, Development and Peace 22,
DOI 10.1007/978-3-319-24496-9_5

The constitution stipulates that elections to these bodies must be conducted every five years, *grama sabha*/ward *sabhas* (citizen's forums) must be constituted for people's participation, and a proportion of seats must be set aside for disadvantaged groups, commonly known as Scheduled Castes/Scheduled Tribes. Such democratic decentralization and participatory planning also encompasses women's empowerment, where one third of seats in the elected bodies and 50 % of the seats in local governing bodies are reserved for women. In addition, 10 % of the total development fund of local bodies is earmarked for projects directly benefiting women.

This chapter provides a brief account of Kerala state's decentralization process and its achievements, particularly in relation to participatory budgeting and women's empowerment. Beginning with a detailed account of participatory planning and budgeting in local government bodies, the chapter describes the process of development as expanding people's capabilities. It discusses how local governments in Kerala are strengthened by devolving functions and finances, and the democratic process deepened by entrusting decision making to elected representatives. Concerted efforts are made to actively involve people in local governance at every stage, and embed women's empowerment and poverty reduction in the system. Expanding capabilities of the poor and marginalized allows them to participate in mainstream local economic development activities. After introducing the local government system and the powers and functions of the *grama sabha*, this chapter explains the participatory planning and budget process with case studies showing women's empowerment. It concludes by charting the elements of success, including the lessons learnt from these reformed structures and processes.

5.2 Local Government in Kerala

Kerala, India's southernmost state, strengthened its local governance system by legislating the Kerala *Panchayat Raj* Act 1994 (for local bodies in rural areas) and the Kerala Municipality Act 1994 (for local bodies in urban areas) (Government of Kerala 1994a, b). The state's powers, functions and financial resources were transferred to these local bodies so they could function as institutions of governance. Planning for economic development and social justice became the responsibility of local governments. About one-third of the Kerala government's development fund was distributed among local governments based on pre-determined criteria. The functional areas of local governments were expanded by transferring agricultural offices, animal husbandry and fisheries offices, health centres (ayurveda, allopathy, homeopathy), schools and *anganawadis* (women and child care centres) to local bodies. A committee system was introduced at all levels

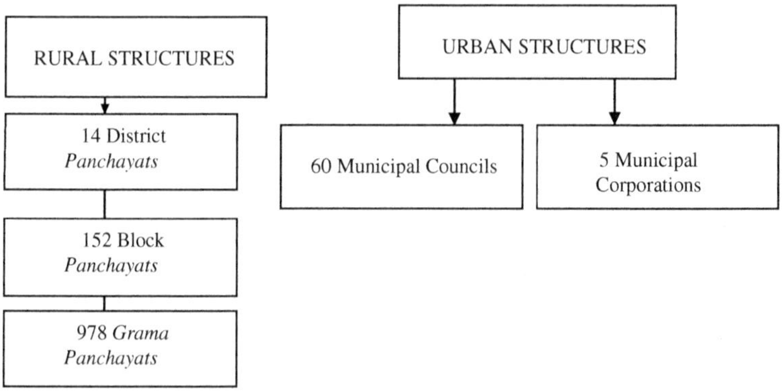

Fig. 5.1 Distribution of LSGIs in Kerala. *Source* The author

within local bodies to expand opportunities for participation. Thus, local institutions,[1] and their officials, under line departments (agriculture, animal husbandry, fisheries, etc.) were transferred to local bodies.

Structure of local self-government institutions: In line with the Constitutional (Seventy Third Amendment) Act, 1992, the Kerala *Panchayat Raj* (KPR) System has a three-tier structure made up of Village, Block and District *panchayats*. Kerala has 1,209 local self-government institutions (LSGIs) (Government of Kerala 2011). Their distribution and a broad structure of the *panchayat raj* system in Kerala are provided in Fig. 5.1.

Grama (village) panchayat: The village *panchayats* have their own sources of income and also receive a reasonable amount as grants and shared taxes. They have also been provided with the major share of planning funds.

Block panchayat: Kerala has 152 Block *panchayats*. Each consists of (a) members directly elected by the people and (b) Presidents of the *grama panchayats*, within each Block *panchayat* area. The President and Vice-President of the block *panchayat* are elected from among the elected members of the concerned village *panchayats*. At present, a block *panchayat* obtains support from two Standing Committees, one on matters of finance and planning and the other on welfare.

Apart from the 24 sectoral functions, the block *panchayats* must pool technical expertise from both governmental and non-governmental agencies at the Block level, to provide technical services to the village *panchayat* to avoid duplication and provide appropriate linkages. Block *panchayats* have no taxation powers. However, they collect fees and avail of loans for financing development programmes. The planning fund is also shared with block *panchayats*.

District panchayat: Kerala has 14 district *panchayats*. The district *panchayat* consists of members directly elected by the people and the presidents of block

[1]Local bodies are elected bodies, whereas local institutions are government departments, which are part of the bureaucratic and technocratic system.

panchayats within the territorial limits of the district *panchayat*. The President and the Vice-President of the district *panchayat* are elected from among the elected members of the concerned block *panchayats*. The district *panchayats* are entrusted with plan integration and mobilization of technical expertise in addition to other sectoral functions.

District *panchayats* have no taxation powers. However, they collect fees and avail of loans for financing development programmes. District *panchayats* can also issue revenue bonds and raise money for financing developmental work. The state government provides grants to district *panchayats*. Furthermore, a portion of the state planning fund is distributed among district *panchayats*.

Committee system: Within the *panchayat*, decisions are based on majority votes by members. At least one *panchayat* meeting must be held every 30 days. To ensure further decentralization within the *panchayat*, standing committees are constituted from among the members. Each *panchayat* member must be part of one standing committee such as Finance, Development and Welfare. To co-ordinate the functioning of the Standing Committees and to avoid conflicting decisions on related issues by different Standing Committees, each *panchayat* has a Steering Committee headed by the president of the *panchayat*.

Grama sabha (Citizens' Forum): The *grama sabha* is the soul of the *panchayat raj* system in the state and its functions and responsibilities have been strengthened further by the KPR Act. In fact, it has become one of the important institutions involved in decentralized planning in the state. The *grama sabha* consists of all persons included in the electoral roll of a ward (constituency) of a *grama panchayat*. The *grama panchayat* member representing the constituency is the Convenor of that *grama sabha*, the meetings of which have to be presided by the president of the *grama panchayat*.

The village *panchayat* must place before the *grama sabha* reports on the developmental programmes of the constituency during the previous year and those proposed to be undertaken during the current year. This includes the expenditure, the annual statement of accounts and the administration of the preceding year. The *grama sabha* must approve the selection of individual beneficiaries for various schemes.

The *grama sabha* is responsible for compiling information and identifying priorities for development plans, helping in their implementation, identifying beneficiaries for various programmes and investigating their eligibility, mobilizing labour and funds for relevant programmes, providing assistance on matters related to public facilities, generating awareness on matters of public interest, organizing cultural activities, providing feedback on various schemes and plans, cooperating with other *grama panchayats* as and where necessary and performing other functions as may be prescribed from time to time. Most *grama sabha* participants are women. This is mainly due to the empowerment of women by means of poverty reduction through the *Kudumbashree* programme which will be discussed in the next section. In other states, where women's empowerment programmes are not as successful, women's participation in local governance has been minimal. In Kerala, women are vigilant and keen to be involved in development activities.

5.3 The Principles of Governance

The decentralization process in Kerala can be better understood by means of the following six principles within the governance process:

Autonomy: Autonomy has three dimensions: administrative, financial and functional. The administrative autonomy of local bodies is guaranteed by limiting state government supervision to obligatory and regulatory functions and making elected bodies—*grama panchayats* (lowest tier in rural areas), block *panchayats* (intermediary tier in rural areas), district *panchayats* (highest tier in rural areas), Municipalities and Corporations—administratively independent units. With the objective of financial autonomy, financial resources are shared, keeping in mind functional responsibilities and as per certain pre-determined criteria, including size of population and area, agricultural patterns and poverty levels based on the nature of employment of the people. These criteria ensure that each local body has a meaningful share considering its requirements, poverty level and potential.

Subsidiarity: This principle governs the division of functions among different tiers of LSGIs. Subsidiarity implies that what can be done best at a particular level should be accomplished at that level and only residual functions should be reserved for higher levels. This principle is strictly followed while transferring institutions under line departments to local bodies. Thirteen departments have transferred institutions and their functionaries to local bodies.

Complementarity: Integration is perhaps the most important task in decentralized planning. The entire system has to work with unity of vision and diversity of means. Overlap and repetition must be avoided while integrating the various functions. Activities at higher administrative levels should thus complement those at lower levels and programmes implemented by all agencies in a given local body should be consistent with local needs and priorities which will, in turn, converge into an integrated local plan.

Uniformity: The principle of uniformity stipulates that norms and criteria for selection of beneficiaries, identification of sites, prioritization of activities and pattern of assistance within a given local body should be the same for a particular programme irrespective of the agency sponsoring it. This helps formulate a standardized criteria for decision making, which could be shared with the public so that transparency can easily be ensured.

Participation of people: In decentralized governance, people's participation is envisaged by empowering them to take their own decisions after they analyse their problems themselves. In fact, local self-government provides an institutional framework for participatory democracy. People's participation is expected at all stages of a development programme including identifying needs, formulating activities and implementing, monitoring and evaluating them.

Transparency and accountability: LSGIs are primarily accountable to the people. This calls for a social audit system to examine the performance of LSGIs. *Grama sabhas* perform this audit by examining the activities of development programmes in the respective areas. Every decision has to be based on certain predetermined norms and criteria that result from social consensus, with the rationale behind each decision being made public.

Based on these principles, Kerala has made democratic decentralization a means of devolving powers and functions to LSGIs on the one hand and introducing decentralized planning on the other.

5.4 Planning from Below

The most important and visible part of Kerala's decentralization is the decentralized planning system, popularly known as the People's Plan (Isaac 1996, 1997, 1999). Extensive participation of the people in every phase of the process is the hallmark of the new system (Isaac/Harilal 1997). All individual beneficiaries are selected by the *grama sabha*, or the Neighbourhood Groups (NHGs) which are further downstream. Thus, the planning process starts from the lowest units in the local governance system. In 1996, the Kerala government initiated the planning process by earmarking one third of the Plan Grant-in-Aid to local bodies annually, for their 'planning for economic development and social justice' (Government of Kerala 1996). Every year, local bodies are asked to prepare their plans (Government of Kerala 2007). Training and capacity building has been an important aspect of this process. Decentralized planning in Kerala has five stages (George/Balan 2011; George/Neunecker 2013).

Stage I—grama sabha: The decentralized planning begins with a special *grama sabha* meeting to understand the needs of people and identify potential areas of intervention. After a brief introductory session, group discussions are conducted and points are reported in plenary sessions. The suggestions of the *sabha* are forwarded to the *grama panchayat*, the lowest tier.

Stage II—development seminar: The second stage in decentralized planning is a Development Seminar at the local body level. The needs, problems and opportunities highlighted in the first stage have to be subjected to analysis given existing resources and materials. Such an exercise results in the publication of a *panchayat*/municipality Development Report. A draft of this report is used as background material for the Development Seminar attended by elected representatives, officials, experts and representatives nominated by *grama sabhas*. The Seminar discusses and suggests strategies and identifies projects to be undertaken for the development of the region. The Development Report is published to ensure wider dissemination of information. All 1,215 local bodies in Kerala have their own published development report.

Stage III—preparation of projects: At this stage, the development strategy is transformed into projects. This is done by Task Forces comprising elected representatives, officials and experts. Working Groups are constituted for important sectors such as Watershed Management; Local Economic Development; Poverty Reduction; Development of Scheduled Castes; Development of Women and Children; Health, Water Supply and Sanitation; Infrastructure and Social Security.

Details on each project include specific objectives, expected cost, resources, contribution from beneficiaries, contribution from the public, mode of execution, mechanism for monitoring and time frame.

Stage IV—preparation of plan document: At this stage, local bodies finalize their plan document. It is not just a compilation of the projects prepared at the ward level. These projects have to be prioritized within resource limits and with a long-term perspective. Unlike in the past, when grant-in-aid was divided ward-wide, allocations are now based on priorities set for the overall development of the region. The elected body democratically decides priority projects to be included in the published Plan Document.

Stage V—plan approval by district planning committee: Every Local Government's Plan has to be approved by the District Planning Committee (DPC), which examines each project of the local body with the assistance of a Technical Advisory Group (TAG) constituted at the Block (intermediate) and district levels. The TAG, consisting of official and non-official experts, examines the technical viability of projects and recommends them for DPC approval. In case of technical imperfections, the local body has to make changes, but the TAG neither rejects nor changes the priority of the project.

Stage VI—implementation, monitoring and evaluation: Each project is implemented under the supervision and control of local bodies. Projects can be implemented by beneficiary committees, the local body, transferred institutions, accredited agencies and through contracts. A Monitoring Committee oversees the progress of the work. The *Grama Sabha* has the right to examine the progress and other details of a project. All individual beneficiaries are selected by the *Grama Sabha* using predetermined criteria provided by the local body. Local bodies do not have the authority to change the priorities set by the *Grama Sabha*. The following section discusses a particular programme emerging from such devolution of responsibility.

5.5 *Kudumbashree*—The Poverty-Reduction Programme for Women's Empowerment

In Malayalam (Kerala's local language), *Kudumbashree* means prosperity for the family and is the name given to a participatory poverty-eradication programme in Kerala. It is a community initiative using women's agency and empowerment with legal backing, implemented as part of the *Panchayati Raj* System (both rural and urban). *Kudumbashree* has an economic base of its own (through micro savings by members) which guarantees autonomy and sustainability. The project is implemented through neighbourhood groups of poor women.

Kudumbashree differs significantly from conventional poverty-alleviation programmes in the identification of the poor in that it undertakes surveys based on nine indicators of risk factors. The risk factors are: (1) poor housing quality; (2) lack of access to drinking water; (3) lack of access to sanitary latrines; (4) number of illiterate adults in the family; (5) single-income households; (6) number of individuals getting barely two meals a day or less; (7) number of children in the family below the age of five; (8) number of cases of alcoholism or drug addiction in the family; and (9) scheduled caste or scheduled tribe status. Households with four out of nine factors are considered poor.

Kudumbashree has a three-tier system. The lowest level are NHGs of 15–40 adult women, one each from identified risk families. Area Development Societies (ADS) are formed at the mid-level by federating the NHGs within a ward of a local body. At the local body level, Community Development Societies (CDS) federate the ADS in the local body. A major function of *Kudumbashree* is to act as an informal bank for the poor. Women pool their savings at the weekly NHG meetings, while the ADS are authorized to open bank accounts to deposit the savings.

Kudumbashree plays an important role in the decentralized planning process. An NHG prepares a micro plan based on surveys and discussions. The concerned ADS then puts together micro plans and prepares a mini plan for the ward. A Monitoring and Advisory Committee is formed under the chairmanship of the elected member of the village *panchayat* or municipality representing that ward. The CDS is co-terminus with the village *panchayat* or municipality and prepares development plans at the local level by consolidating the plans prepared by the ADS. The case study below further elaborates the issues raised, particularly pertaining to women's empowerment.

5.6 A Case Study of Pazhayannur Village *Panchayat*

Pazhayannur *Grama Panchayat* was established in 1954. With an area of 59.03 km^2, Pazhayannur falls in Thrissur district under the Alathur Lok Sabha constituency and the Chelakkara Assembly constituency. Its population of 43,325 (men: 21,005 and women: 22,320) boasts a literacy rate of 85 %.

Administration: The *grama panchayat* is governed by the *panchayat* committee headed by a president. The committee has 22 members elected from 22 wards; of these, ten members are women (50 % seats are reserved for women in local governments in Kerala). Four Standing Committees have been formed by dividing the 22 members into four groups: (1) Finance; (2) Development; (3) Welfare and (4) Health and Education.

Impact of devolution of finance: As noted earlier, from 1996–1997, the village financial position improved substantially as a result of the introduction of the 'People's Campaign for the Ninth Plan', which was initiated to strengthen local government. One-third of the total Development Fund of the state government is distributed among local governments for 'planning for economic development and social justice'. In 1990–1991, the budget was Rs. 1,599,000, increasing marginally each year.

The income of the Pazhayannur *grama panchayat* rose from Rs. 1,599,000 (Rs. 1.6 million) in 1991 to Rs. 3,231,000 (3.2 million) in 1996–1997. The flow of funds to Pazhayannur increased substantially following decentralization and devolution of finance. The *grama panchayat*'s financial resources increased to Rs. 17,124,080 (17.1 million) in 1997–1998, an increase of more than five times from the previous year. It stayed around this range till 2002–2003, and began rising steadily after that, with an increase of almost Rs. 10 million each year. In 2011–2012, the budget was Rs. 100,570,509. Most of the *panchayat*'s financial resources came from the state government and from other agencies with about 22 % of the total funds being mobilized by the local body through village-level tax collection.

Transferred institutions: Along with financial devolution, institutions are also transferred to local bodies in Kerala. Several institutions were transferred to Pazhayannur *grama panchayat* including Krishi Bhavan (agricultural centre), veterinary dispensary, primary health centre, Ayurveda dispensary, homeo dispensary, lower primary, upper primary and high schools, and a diary development office. In addition seven health sub-centres and 38 *anganwadis* were also transferred.

Working group for participatory budgeting and planning: Ten working groups were constituted in the *panchayat* for participatory budgeting and planning. The Chairperson of the working group is an elected representative while the Convener is the official head of the concerned sector. These working groups consolidate discussions at the *grama sabha* and transform the aspirations of the people into implementable projects.

5.6.1 Achievements

Participatory planning projects: A project is the smallest unit of implementation in the Plan. The Pazhayannur *grama panchayat* had 116 projects in 2013–2014. These included all sectors and targeted the poor and marginalized in society, helping them to build their capabilities and allowing them to live with dignity.

Pazhayannur *grama panchayat* has 16 Agriculture Production Units under *Kudumbashree*. Each unit cultivates vegetables in an area of about an acre and has 10–15 persons, most of them women, as self-employed members. The units have local marketing outlets and sell their products mainly among the people in the area. Excess output is sold to the Vegetable and Fruit Promotion Council Keralam (VFPCK), a state organization for the promotion of agriculture and horticulture. Members own cows and produce organic vegetables using dung as fertilizer. They earn Rs. 300 to Rs. 350 per day.

The *panchayat* has two garment-making units that stitch school uniforms. The other units are engaged in notebook binding, assembling umbrellas and manufacturing bags for ladies. Some of the units market their products through schools as their production is based on demand from schools in the surrounding area. On average, members of these units earn Rs. 4,000 to Rs. 5,000 per month.

Box 5.1: Case Study 1: Initiatives Directly Benefiting Women

The Pazhayannur village *panchayat* has several activities aimed at empowering women. Ten per cent of the local government's Development Fund is earmarked for activities directly contributing to women's empowerment. In 2013–2014, seven development projects were implemented at a cost of Rs. 3 million. These included Joint Farming, Women's Labour Bank (*Mahila Kisan Sasaktikaran Paryojana* or MKSP), Housing for Widows, *Kudumbashree* Goat Farming, Support for Women and Child Care Workers, Financial Support for Scheduled Caste Women and Food for *Anganwadis* (women and child care centres). Under the financial support programme for SC weddings, Rs. 1.8 million was distributed (Rs. 50,000 per head). Under the Housing for Widows initiative, ten women were given Rs. 200,000 each. Besides such disbursement of funds, the local government also organized seminars and discussions to sensitize people about women's empowerment.

Box 5.2: Case Study 2: From Weaving Threads to Contesting Panchayat Elections

Surya is a self-help group under the *Kudumbashree* programme, producing thread from cotton. Women workers employed in Surya were provided training which was organized by the *panchayat* and conducted by the Khadi and Village Industries Board. The total investment for the unit was Rs. 200,000, 50 % of which was subsidized by the Khadi; the other 50 % was a bank loan. On average, Surya makes 400 cotton units in a month. Each worker takes home wages of Rs. 1,000 per month and Rs. 2,400 as a share of income. Marketing is not a problem as their entire production is purchased by the Khadi and Village Industries Board. Further, the Board pays each unit member Rs. 1,200 for their children's needs. The machines used to make the thread are installed in the homes of members. They run on electricity and women manage the machine along with their regular household activities.

Sujatha Narayanankutty (34) is a member of the Surya Cotton Thread Making Unit in Pazhayannur *grama panchayat*. She has two school-going children. Her husband is a labourer. She has been working in the unit for the past six years, and earns about Rs. 4,000 per month. It helped her build a good relationship within the community, which has increased her self-esteem. Sujatha's association with the programme also brought about banking literacy, which contributed substantially to her saving habits and orientation towards entrepreneurship. She contested in the *panchayat* elections, and although she did not win, she gained popularity in the local community. Now she feels that her social status has increased.

Box 5.3: Case Study 3: Tailor Becomes *Panchayat* President

Devi Durga *Kudumbashree* Tailoring Unit, Neernamukku, was established in 1999 in Ward 1 of the Pazhayannur *grama panchayat*. It has 17 members and purchased 1,340 sq. ft. of land at a cost of Rs. 14,000 and constructed a building with a loan. Of the 17 members, five work in the Tailoring Unit to manufacture clothing for women and children. They also train other women. Each member earns Rs. 3,000 to Rs. 3,500 per month. To meet capital costs, they procured a loan of Rs. 400,000 from the State Bank of Travancore, a nationalized bank.

Rema Radhakrishanan, a member of the Devi Durga Tailoring Unit, is the former *panchayat* president. Her family has about 1,600 sq. ft. of land and a small house. She has two college-going children and her husband is a daily wage labourer. She has been able to support her children with sustainable income from the tailoring unit. Rema became popular in the local community after joining *Kudumbashree* and eventually was elected as the president of the *panchayat*. She says that her leadership qualities, capabilities and courage were acquired from her experience as a *Kudumbashree* member.

Box 5.4: Case Study 4: Household Entrepreneurship to Leadership

Santhakumary (43) and her husband produce and market semi-processed food, specifically for preparing popular local breakfast items such as *puttu*, *dosa* and *idli*. She credits her experience in community involvement at Krishna *Kudumbashree*, and believes that she could become a *panchayat* member only because of the exposure she received as a member of *Kudumbashree*. She took a loan of Rs. 259,000 from the Service Cooperative Society to cover the capital cost of the four machines needed for her business. Santhakumary employs ten people who process about 100 kg of rice and produce 300 packets for sale every day. Interestingly, Santhakumary's husband, formerly a house painter, now markets the products. She credits all her achievements to her *Kudumbashree* affiliation: becoming a *panchayat* member, gaining banking literacy, earning regular income for her family, developing contacts with the local community and improving her social status.

5.7 Conclusion: Lessons from Participatory Planning and Budgeting in Kerala

The development of a society depends, to a great extent, on the development of social capital. Social capital is the networking that helps create links which, in turn, forge rules, conventions and norms governing the development process. When a society undergoes change, certain institutional factors that have a spiralling effect on deepening democracy could gain ground if the unleashed energy of people is consciously consolidated. Decentralization in Kerala led to the conscientization of the masses coupled with the support of numerous activists who worked beyond the interests of their political affiliation/inclination (Friere 1970). Democratic principles gained ground and decisions relating to the public had to be taken democratically in line with social justice principles. As a result, ordinary people, including women, have come to understand the importance of human development and, to some extent, the concept of sustainable development. In sum, the net result of decentralization was sustainable human development and women's empowerment in Kerala.

Our attempts at evaluating the performance of this process, by means of a rapid appraisal, identified the following elements of success: (i) empowerment of women; (ii) local-level democratic self-governance; (iii) efficacy; (iv) attitude change among staff; (v) participation; (vi) equitable prioritization.

Empowerment of women: Perhaps the most significant aspect of participatory planning and budgeting is the participation of women in local governance and the development process, leading to a substantial transformation in the social development of the people in Kerala. When women became leaders in the governing bodies of local governments, priority automatically shifted towards improving the capabilities of the people. Women leaders and activists took the initiative to provide health care for women during pregnancy, children were taken care of at *anganawadis* (women and child care centres), children were given education in schools and supported with special tuition facilities, and special nutritional support was arranged for adolescent girls. Through *Kudumbashree* and its participatory approach to women's leadership, the institutional framework was built to provide for the sensitization and organization of development for poor women. Even the beneficiaries are identified by the women themselves.

Local-level democratic self-governance: Kerala has experienced a significant improvement in economic development and social change with the strengthening of local governments. Formerly, these development processes were administered by officials in government departments with no involvement of either the people or their representatives. Local governments in Kerala, both rural and urban, were in existence but their involvement was limited as they lacked power, function and financial resources. With the introduction of participatory planning and budgeting along with the devolution of functions and finances, the quality of activities undertaken under development projects has improved substantially. This structural change also affected the nature of development activities. When the role of the local governments changed from merely implementation to participatory planning and budgeting, local bodies began to formulate ideas, identify projects and implement, monitor and evaluate them.

Efficacy: Time-bound implementation of projects is an important achievement; it also reduces expenses. Every year, each local body has to prepare a planning document that contains the details of projects to be implemented in the year. In addition to budgeting for the current year, local bodies also prepare a plan for the ensuing year that can be revised later. Further, they prepare a five-year plan document that contains major development activities to be undertaken during the next five years.

Attitudinal change among staff: When local governments became autonomous, the efficiency level of officials improved. With empowered elected representatives governing, officials began to be recognized by the governing body. Officers work to their full potential when their professional performance is appreciated, acknowledged and supported. For example, in health centres, lack of resources resulted in a shortage of medicine and doctors were merely handing out prescriptions. When the health centre became part of a local government governed by elected local representatives, there were fewer delays. Now, in case of a shortage of medicine, medical officers can instantly inform the local government and decisions for purchasing medicine can be taken immediately. Doctors are happier as patients obtain treatment, not just prescriptions.

Level of participation: Before decentralization, people's participation in development activities was minimal, limited to electing representatives to local governments, which was an irregular process. Now, elections to local governments are conducted every five years. Fifty per cent of the seats in the governing body are reserved for women. Even the governance of the local body is decentralized by dividing responsibilities among standing committees within it. Every elected representative must be a member of one (and only one) such standing committee. Every elected representative represents one *Grama Sabha* in the governing council.

The participatory planning and budgeting process opens up immense opportunities for people's and women's involvement in development activities. Ideas for development activities germinate from the *Grama Sabha* and are transformed into implementable projects with help from the people who support the local governments voluntarily. Identified projects are implemented with the involvement of the people, often through beneficiary committees.

Equitable prioritization: The change in the process of decision making is another key achievement. Before decentralization, every activity was centrally identified and the role of local governments was limited to implementation. Currently, all decisions relating to development activities are made by the local bodies, who decide on the activities to be undertaken, prioritize these activities and identify beneficiaries, including individuals. All these processes have an underlying fundamental question of equity. With the introduction of participatory planning and budgeting, attention has focused on lower-income groups. Now, the first priority is accorded to the destitute; each local government is mandated to support the destitute and provide them with food, accommodation, clothing and medicine.

Although it is premature to evaluate the experimental decentralized planning, considered a pioneering attempt globally, certain impressions may be gathered from the experience. The achievements registered so far have attracted worldwide attention and acceptance as a new methodology in decentralized planning. Justice Krishna Iyer, renowned social activist and thinker, noted: "The people's campaign marks the beginning of a social revolution in Kerala. The failure of democracy in India consists in the failure to share power with the people. In Kerala today political power is in the hands of the people" (Iyer 1998).

In summary, the overarching achievement of Kerala's decentralized governance is its acceptance by all sections of society, including various political parties. In a multi-party parliamentary democracy, it is rare for all political parties to support decentralization and participatory planning. In fact, they are competing to strengthen local governments. Such an approach is undoubtedly beneficial to the people of Kerala and could prove equally important in other Indian states and other countries as a model for achieving equitable social development along with women's empowerment.

References

Centre for Development Studies (CDS), 1980: *Some Notes on Possibilities of Decentralised Development in Kerala.* Working paper 106 (Thiruvananthapuram: CDS).

George, Sunny; Balan, P.P., 2011: "People's Participation in Development Planning in Kerala", Paper for India Urban Conference, Mysore, 17–20 November.

George, Sunny; Neunecker, Martina, 2013: "Democratic Decentralisation and Participatory Budgeting: The Kerala Experience", in: Sintomer, Yves; Traub-Merz, Rudolf; Zhang, Junhua (Eds.): *Participatory Budgeting in Asia and Europe* (London: Palgrave Macmillan): 66–85.

Government of India, 1992a: *73rd Amendments to the Constitution of India* (New Delhi: Government of India).

Government of India, 1992b: *74th Amendments to the Constitution of India* (New Delhi: Government of India).

Government of Kerala, 1994a: *Kerala Municipality Act 1994* (Thiruvananthapuram: Government of Kerala).

Government of Kerala, 1994b: *Kerala Panchayat Raj Act 1994* (Thiruvananthapuram: Government of Kerala).

Government of Kerala, 1996: *People's Campaign for 9th Plan in Kerala: An Approach Paper* (Thiruvananthapuram: Kerala State Planning Board).

Government of Kerala, 2007: *Guidelines for the Preparation of Annual Plan (2007–2008) and Eleventh Five Year Plan (2007–2012) by Local Governments* (Thiruvananthapuram: Department of Local Self Government).

Government of Kerala, 2011: *Panchayat Vikasana Guide* (Thiruvananthapuram: Department of Local Self Government).

Isaac, Thomas; Harilal, K.N, 1997: "Planning for Empowerment: People's Campaign for Decentralised Planning in Kerala", in: *Economic and Political Weekly*, 32 (1–2): 53–58.

Isaac, Thomas T.M, 1996: *People's Campaign for 9th Plan: An Approach Paper* (Thiruvananthapuram: Kerala State Planning Board).

Isaac, Thomas T.M, 1997: *Janakiyasoothranam: SidhanthavumPrayogavum* (in Malayalam) (Thiruvananthapuram: Kerala State Planning Board).

Isaac, Thomas T.M, 1999: *People's Planning: Towards a Handbook* (Thiruvananthapuram: Kerala State Planning Board).

Iyer, Krishna, 1998: "Beginning of Revolution", in: *Mathrubhumi*, 17 August.

Part II
Implementing Gender Responsive Budgeting Tools for Gender Equality and Social Justice

Giving the participants a taste of Malaysian culture and arts by staff of the Seberang Perai Municipal Council. The photo was provided by PWDC who granted permission to include it in this volume

Chapter 6
Integrating Gender Equality, Women's Rights and Participation in the Budget Process: A Survey of Entry Points and Practical Examples

Elisabeth Klatzer

Abstract This chapter will demonstrate how GRB can be integrated into the regular budgeting process using a framework of the four stages of the budget cycle. Identifying entry points for GRB at each of these stages, the paper then discusses possibilities for GRB work for both government and civil society actors. GRB instruments and case studies from around the world are used to concretize the various processes. A key to the success of GRB is to go beyond technicalities and procedures and work towards transforming institutions and processes.

Keywords Institutionalizing GRB · Budget cycle · Gender auditing · Gender planning · Budget processes

6.1 Introduction

Practical Gender Responsive Budgeting (GRB) work in regular budget processes can be implemented using a broad range of approaches that basically fall into two categories: (i) undertaking selective GRB work related to selected programmes, policies or activities or at selected, isolated points in the budget process, which remains a piecemeal approach or (ii) a systematic integration of gender equality, women's rights and participation throughout the planning and budgeting process which attempts to work towards full institutionalization and thus towards transforming public policies and processes. This chapter focuses on the latter and aims to highlight tools and approaches to how GRB can be integrated into the regular budgeting process. Practical experiences show that GRB is more promising if actors follow a gender responsive approach in standard processes and activities. It is even more effective if gender equality and women's rights advocates from within and

Elisabeth Klatzer, Feminist Economist, Researcher, Activist and Consultant, Vienna, Austria. Email: elisabeth.klatzer@gmx.net.

C. Ng (ed.), *Gender Responsive and Participatory Budgeting*,
SpringerBriefs in Environment, Security, Development and Peace 22,
DOI 10.1007/978-3-319-24496-9_6

outside public administration are closely involved in these processes. Even more so, GRB can only be successful if it combines policy change processes from within government with participatory processes, not only 'by invitation', but especially 'by irruption' (see Ng, this volume). In fact, the concept of GRB inherently implies ample gender responsive participatory approaches to budgeting if fully implemented as a transformative approach. Otherwise it is in danger to remain a government cosmetic approach leading to marginal changes at best and leaving underlying power structures, and access as well as ownership of resources, unchanged. This is an aspect that is too often overlooked in pragmatic GRB work by public administration.

To highlight how GRB can be integrated into the regular budgeting process, a simple framework of the four stages of the budget cycle is presented as a starting point (Sharp 2003: 70). Subsequently, entry points for GRB are identified at each of these stages. Possibilities for GRB work for both government and from below by civil society actors are presented. A series of GRB instruments to be used throughout the budget cycle are presented. To make this approach easily accessible for GRB practitioners, examples of actual GRB work at different stages (case studies from across the world) are used to illustrate how it can work in practice. A major challenge is to go beyond technical procedures and work towards full transformative institutionalization.

6.2 A Framework for Implementing GRB in the Budgeting Process

This section presents a framework for integrating GRB throughout the different stages of the budgeting process. Despite country-specific variations, the budgeting process can be described in four stages (see Sharp 2003):

1. budget formulation (or preparation);
2. budget adoption (or enactment);
3. budget execution and implementation; and
4. budget control: audit and evaluation.

In many countries, the annual budgeting process is preceded by a multi-annual budget planning process, often called multi-annual budget framework, representing an envelope for annual budgeting, mostly by defining the upper limit for annual budget expenditure.

Different opportunities exist for integrating GRB into the budgeting process in each of the four stages (in the context of institutionalizing it). The following section describes the main activities at the different stages of the budgeting process and identifies possibilities for GRB activities at each stage (Sharp 2003: 70; Hadžiahmetović et al. 2013). The institutionalization of a gender equality and women's rights perspective (Elson/Cagatay 2000; Elson 2006) into all stages of the budget cycle is a crucial—absolutely necessary, but not sufficient—prerequisite for achieving a more systematic and transformative implementation of GRB in the budgeting process.

6.3 Integrating GRB and Participatory Elements at the Budget Formulation Stage

Budget formulation involves a series of activities to plan and prepare the budget. At this stage, important opportunities can be availed to ensure the systematic integration of gender equality and women's rights issues. The main activities typically undertaken at this stage of the budgeting process are:

- economic and budget revenue forecasts, as a basis for budget projections by the Ministry of Finance;
- overall objectives and priorities determined by the government;
- budget call or circular issued by the Ministry or Department of Finance;
- sector priorities established by sector departments and ministries;
- output costs estimated;
- budget requests and draft sector budgets prepared by sector ministries and other public institutions;
- budget allocations negotiated by the Ministry or Department of Finance with the sector ministries; and
- a draft budget prepared by the Ministry or Department of Finance, approved by the government and sent to the Parliament or Council.

All these activities offer potential for the integration of GRB. The main possible entry points at this stage include:

- inclusion of gender equality objectives in the highest level objectives, priorities and gender sensitive formulation;
- formulation of activities to implement gender equality objectives and appropriate indicators to measure performance towards achieving gender equality determined;
- inclusion of a request for all line ministries/departments in the budget call circular to integrate GRB into their budget submissions;
- information on the gender impact of programmes in relation to budget items by sector ministries or departments in their budget material submission. For performance budgeting, this would involve specification of gender responsive objectives, activities and measures as well as milestones and indicators;
- gender budget analyses of the main programmes and policies (*ex ante* and *ex post*) by sector ministries or departments as part of budget material; and
- gender budget statements or other forms of documentation of GRB work as part of budget material or as separate reports.

Budget preparations take place inside the government, with the Finance Ministry or Department, line ministries or sectoral departments, the government and top public officials as the key actors. As such, they are responsible for integrating gender equality and women's rights issues. However, their work will be enhanced and improved if the process allows for participation of and contributions from gender equality advocates from within and outside public administration, participation 'by invitation'.

Academics and civil society could apply different methods to promote GRB activities at this stage of budgeting. Researchers could use gender sensitive models to prepare forecasts for budget parameters, present evidence of existing gender inequalities and provide gender sensitive research as a basis for setting high level objectives and priorities. As regards participation 'by irruption', time of budget preparations is an important phase and offers opportunities for mobilization around key issues of importance to achieve gender equality and women's rights, as well as for lobbying and advocacy with different government institutions, especially sector ministries, to include gender-related priorities in policy planning and budgeting and allocate sufficient funds to activities important to women and gender equality. Civil society organizations (CSOs) can influence the process in many ways from outside and draw the attention of administrators and politicians to issues of importance to promote gender equality and social justice.

6.3.1 Examples of GRB Work at the Budget Formulation Stage

The budget call circular is an official notice issued by the Department or Ministry of Finance at the beginning of budget preparations. It contains instructions to other government departments, ministries and agencies on how they must submit their budget requests for the coming budget period. It often contains a very specific format for budget submissions. If governments take GRB seriously, this must be a key entry point for guiding ministries on including gender perspectives in budget preparations and highlighting the importance of GRB for the government.

Many GRB initiatives focus on integrating GRB in the budget call circular. This integration can have different forms as illustrated by the examples in the following section. These are only intended to be illustrative as any practical work will have to adapt the gender responsive instructions to the form and purpose of the specific budget call circulars used. The way that the budget call circular raises GRB issues and the depth and scope of requirements for public institutions to integrate it in budget preparations can be a good indicator of the importance attached to it by the government.

6.3.1.1 Former Yugoslav Republic of Macedonia

The budget call circular of the Former Yugoslav Republic of Macedonia requires line ministries to formulate gender indicators in the framework of their programmes and to submit them with the budget material (UNIFEM 2007).

6.3.1.2 Pakistan

In Pakistan, the budget call circular required—at least while GRB efforts were ongoing—all relevant performance indicators to be sex-disaggregated and gender-related indicators to be included wherever relevant. Ministries and departments have to include gender components when describing their missions, visions, goals and activities. The following extracts from the 2007–2008 federal budget call circular highlight the rather detailed instructions (Mahbub/Budlender 2007: 7):

Box 6.1: Extracts from Pakistani Budget Call Circular. Source: Mahbub and Budlender (2007: 7)

Ministerial policy objectives

Indicate gender-related objectives in the Ministry's own policy as well as objectives relevant to the sector from general policies.

Key output indicators

Indicate the required disaggregation, such as sex/gender and age group for services delivered to individuals.

Input indicators

Indicate the relevant disaggregations. In particular, indicate the current gender/sex breakdown of both qualified and unqualified staff.

Mission statement and functions of the Ministry

Specify where and how functions contribute to the achievement of gender equity.

Objectives statement of the Ministry

Disaggregate targets and performance measures where appropriate. In particular, indicate sex/gender disaggregation wherever possible. Include specific performance measures related to gender equity.

Key output indicators in the medium term

Disaggregate key output indicators wherever relevant. In particular, disaggregate by sex/gender wherever possible. Include specific output indicators related to gender equity.

6.3.1.3 Austria

Instructions in the budget call circular at the federal level varied over the years along the process of GRB implementation. For the 2005 budget, which was during the initial stages of GRB implementation, the budget call circular required each ministry to include results of two pilot gender budget analyses in draft budget submissions. The results were to be included in a specific chapter on 'Gender Aspects of the Budget' as explanatory notes to the draft budget document.

With full implementation of budget reform towards performance-based budgeting and a firm intention of achieving equality between women and men in budgeting in 2013, the budget call circular includes instructions to specify gender equality objectives, measures and indicators in the context of performance budgeting. Each budget chapter—which corresponds roughly to the main policy areas or ministries—has to identify a few overall outcome objectives with a maximum of five per chapter (Steger 2011: 8). At least one of these overall outcome objectives for each chapter has to directly address gender equality. Figure 6.1 illustrates the form that all ministries have to fill for the overall outcome objectives, including gender equality objectives.

Ministries were instructed to specify activities and outputs contributing to reaching outcome objectives. This too had to include gender equality considerations. The form used in the budget instructions is depicted in Fig. 6.2.

Ministries also need to specify milestones and indicators to measure success and provide a base line of current status (see grey column in Fig. 6.2). Furthermore, ministries are instructed to report on developments from previous budget statements and list recommendations for the Court of Auditors and their specific responses. All this applies to gender equality objectives, activities and outputs as well. As such, the gender equality perspective is firmly integrated in the budget preparation process at the federal level in Austria and the call circular gives clear instructions on how to prepare this aspect for budget submissions.

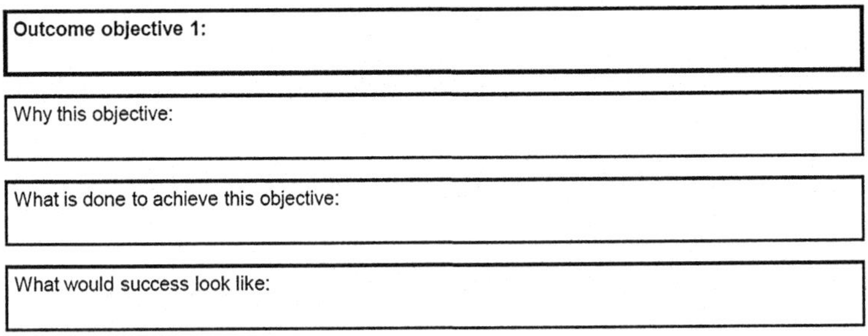

Fig. 6.1 Austria: Form to specify outcome objectives. *Source* Steger (2011)

Activities/Outputs *(max. five including gender activity/activities)*

Contribution to outcome objective/s no.	What is done to achieve the outcome objectives? Activities/Outputs:	What does success look like? Milestones/Indicators for n+1	Current status as indicated in the most recent performance report

Comments on activities/outputs of the preceding budget statement, which are no longer listed in the present budget statement

Recommendation of the Court of Audit

Response of the ministry

Fig. 6.2 Austria: Form to specify activities and outputs. *Source* Steger (2011)

6.3.1.4 Uganda

In Uganda, the central government adopted GRB under the term 'Gender and Equity Budgeting', including other inequality dimensions besides gender. The 2005–2006 budget call circular included the following (Budlender 2007: 4):

> Government is committed to mainstreaming gender and equity objectives in the planning and budgeting process. This involves articulation of the needs of special groups including women, orphans, elderly, youth and the persons with disabilities and integration of the respective interventions with the existing programmes as part of the budgeting process. The guidelines have been incorporated in the Terms of Reference for the Sector Working Groups (SWGs). You are therefore requested to identify these requirements and ensure that the cost implications are integrated into the budget estimates for the next financial year 2005/2006 and the medium term.

Over the years, the GRB approach has been further developed in Uganda. The First Budget Call Circular for financial year (FY) 2014–2015 includes more specific instructions (Ministry of Finance, Planning and Economic Development 2013: 14f) (emphasis retained from original):

Box 6.2: Mainstreaming of Gender and Equity, HIV/AIDS and Environmental Concerns in the Budget. Source: Ministry of Finance, Planning and Economic Development (2013: 14f)

Over the years, Sectors have been requested to articulate Gender and Equity issues, in their Budget Framework Papers (BFP). Analysis of BFP submissions over the years reveals that most sectors just put blanket/general statements that they are addressing the issues. Therefore, effective next FY 2014/2015, Sectors should clearly make budget statements in their BFPs highlighting the Gender and Equity issues, and the actions proposed to address these issues through the budget. The budget statement should spell out what the sector wants to achieve, the issues of concern, and the actual selected interventions with proposed budget allocation.

Example: *Gender aware statement for Education*

Objective: Enhancing gender equality in Uganda

Issue of concern: High school drop-out rates especially for the girl child

Proposed interventions:

(i) Girl Child Scholarships for the needy;
(ii) Recruitment of senior women teachers to counsel girls;
(iii) Construction of the requisite sanitation facilities in schools

Budget Allocation: Ug shs 3 billion (could have specific allocations to each activity)

Performance indicators:

To facilitate the budgeting for these issues, a special window has been created in the Output Budgeting Tool (OBT) as a measure for each vote to demonstrate the specific interventions as well as the budget that has specifically been earmarked for these cross cutting issues.

These examples of budget call circulars including requirements on gender responsiveness also involve the preparation of gender budget statements with gender information. Different forms of gender budget statements are dealt with in greater detail in the next section on integrating GRB at the budget approval stage.

Civil society and broader participation activities at the budget preparation stage can be wide-ranging. These mainly intend to highlight the need for more equitable budget preparations and advocate for more resources for budget items of particular importance to women. Examples are the long-standing work of the Women's Budget Group in the United Kingdom, or the efforts of grassroots mobilization and focused advocacy work by CSOs in the context of the civil society Women's Budget in South Africa.

6.4 Integrating GRB and Participatory Elements at the Budget Approval Stage

The budget is approved in the Parliament or the Council. This stage of the budgeting process normally garners a lot of media attention and is thus an opportunity to highlight GRB issues by both government and civil society. The main activities typically conducted at this stage of the budgeting process are:

- budget presentation to parliament by the Minister of Finance, governor or mayor;
- budget speech of the Minister of Finance;

- budget debates in the plenary and in the respective commissions in Parliament/Council;
- information provided to the media and other stakeholders;
- changes and adjustments to budget allocations;
- draft budget accepted or, in very rare instances, rejected.

The activities at this stage offer potential entry points for integrating GRB. Some of the main possible entry points at this stage are:

- inclusion of information on gender issues in material presented to the parliament by the government;
- specific focus on GRB and gender equality incorporated in the budget speech by the Finance Minister;
- hearings involving civil society and independent researchers organized on budget impact on gender equality during the budget debate;
- discussions between parliamentarians and civil society on the gender equality priorities to be raised in the budget debate;
- changes to budget allocations, e.g., to better meet the needs of women and ensure sufficient funding for gender equality objectives by the Parliament;
- gender equality and women's rights advocates mobilize around priority budget issues and/or lobby for increases in budget allocations for items of particular importance to women and gender equality;
- briefings on the impact of gender equality on budgets and other issues by civil society actors;
- an alternative gender responsive budget proposal, e.g. with an explicit agenda promoting gender equality and women's rights (as an advocacy tool), presented by civil society and/or researchers;
- public relations work to inform the broader public about the impact of the budget on women and men and the importance of GRB;
- public scrutiny and debate on gender equality in the budget; and
- cooperation with the media and other information providers to provide them with relevant information and enhance their reporting on the gender aspects of the budget.

Key players at this stage are parliamentarians, councillors and the Finance Minister, governor or mayor. However, CSOs, the media and researchers, and especially the community at large have ample room to create or use spaces to mobilize and participate actively at this stage and promote increased focus on gender equality and women's rights in the budget and lobby for changes in expenditure allocations. Public attention on budgeting tends to be high at this stage, which can be used to highlight gender equality and women's rights issues within the context of budget policies. It is up to Parliaments and Councils to open formal and invited spaces of active participation for gender advocates at the time of budget debates and allow for processes of gender equitable participation of citizens in decision-making processes about public budgets. However, community mobilizations can create independent spaces.

6.4.1 Examples of GRB Work at the Budget Adoption Stage

A gender budget statement (GBS), a key tool of GRB work, helps showcase the gender responsiveness of the draft budget at the time of its presentation to the parliament or council. A GBS aims to demonstrate the expected implications of budgets in addressing issues of gender inequality (Elson 1999: 7). It is a statement or report from the government on how gender issues are addressed by budgetary policies, expenditure and revenue. It is an accountability document produced by the government to present its efforts towards gender equality (Budlender/Sharp 1998: 50). It can present the results of GRB analysis and put forward recommendations on further changes to objectives, activities and measurement indicators. The form and content of such GBS are a good indicator about the transformative potential of the GRB activities: e.g. technocratic counting exercises are not very promising, while an account of key challenges in inequalities and ways to overcome it by means of policies and budgets would open more worthwhile transformation paths.

In India, the Finance Ministry introduced a GBS for the 2005–2006 Union Budget that identified 'demands for grants' (proposed allocations) that would substantially benefit women in ten departments. For the 2006–2007 Union Budget, the GBS was expanded to 24 demands for grants from 18 ministries. Each ministry identified two categories:

- allocations that were 100 % targeted at women or girls and
- allocations of which at least 30 % was targeted at women or girls.

For each identified allocation, the following were included: the amount allocated in the previous year, the revised amount for the previous year (the amount likely to be actually spent) and the budget for the coming year.

France's GBS, called the Yellow Budget Paper on Women's Rights and Gender Equality, was introduced in 1999. Members of Parliament decided that each year, along with the Finance Bill, the government must table a document showing what it was doing to promote women's rights and gender equality. France also has Yellow Budget Papers on other issues that are the responsibility of more than one ministerial department.

For the gender Yellow Budget Paper, each department is required to:

- specify all programmes and actions targeted at women or girls;
- specify all actions taken to promote gender equality;
- describe policy on gender equality; and
- present the indicators used to measure its contribution.

Analysis of earlier Yellow Budget Papers showed that gender specific allocations accounted for only about 40 million Euros out of a total budget of 260 billion Euros. These results illustrate the importance of including mainstream programmes that promote gender equality in the yellow budget paper.

Since 2003, the government of Gauteng Province in South Africa has required every department to produce a GBS as part of its main budget document. As in many other countries, the statement does not describe a separate budget for women and gender equality; instead, it describes sub-programmes that are already part of the standard budget for the department and their specific relevance for women and gender equality. The GBS uses a prescribed matrix format wherein most columns use similar terms to those used in the main part of the budget, such as outcomes, outputs and indicators. However, it also has a column labelled 'gender issues' where the department must describe the issue that a particular sub-programme is attempting to address.

Departments are asked to prepare their GBS in five parts:

- outcomes and outputs of the three largest sub-programmes and their implications for gender equality;
- outcomes and outputs that specifically target women and girls;
- outcomes and outputs that will benefit women/promote gender equality;
- outcomes and outputs that will benefit women employees within the Gauteng Provincial Government; and
- the number of women and men government employees by level.

The City of Vienna, Austria publishes an annual GBS that is part of the budget material presented to the Municipal Council. In addition, a performance report evaluating implementation measures is included in the annual audit report for the preceding year. All administrative units have to provide information within a simple general format. The Table 6.1 shows the general format of the Viennese GBS and provides a few examples of the information that departments have to provide for different budget items.

In addition, the GBS includes statistical information on the development of gender equality and summarizes developments by means of a gender equality index developed in Vienna. Sex-disaggregated data on key issues of importance to gender equality, e.g. the labour market, income, education, child care, etc. is presented in the GBS as well.

The Government of Pakistan produced its first GBS in 2008–2009 as part of the Strengthening Poverty Reduction Strategy (PRS) Monitoring Project aimed at the promotion of gender equality (Government of Pakistan 2009: 6). The GBS covers select pilot programmes, the Benazir Income Support Programme and other projects selected by the ministries of education, health and population welfare. Initially, programmes with considerable gender implications were selected. Box 6.3 is an excerpt from the Pakistan GBS 2008–2009.

Table 6.1 Gender budget statement of Vienna, Austria (examples)

Department	Budget item	Gender specific objective	Planned measures or projects	Success indicator	Planned users (female/male) in %
MA 42 (City Parks and Gardens)	8150	Gender sensitive, cross-generational measures when designing parks	10. water playground	Number of parks	49/51
			11. park of N. estate, etc.		
MA 5 Finance and Statistics	0210	To comprehensively represent the life realities of women and men	Increase in gender specific information in regular municipal publications. Data and analysis on different areas with specific attention to gender differences	Increase in the share of data presented by women/men	50/50
			Online resource on gender equality indicators	Increase in the share of analyses that include relevant information on men and women users and distribution	
			Evaluation of proportion of sex disaggregation of online data	Information as a basis for evaluation of online data presentation	
MA 5	7822 (ZIT Technology Agency Vienna)	Increase the share of women in business, research and development; increase attention to gender aspects in the development of innovations	Projects led by women to receive an additional bonus. Attention to the equal representation of women and men in project assessment groups	Increased share of projects led by women	20/80
			All supported research projects will be evaluated with regards to the integration of gender issues in project planning	Increased participation of women in research and development	30/70
				Increased attention to the specific impact of product developments on women	

Source City of Vienna, Gender Budgeting, in annexe to the 2012 draft budget

Box 6.3: Pakistan's Gender Budget Statement for 2008–2009. Source: Government of Pakistan (2009)

Gender Budget Statement: Education

Role of Federal Government:

Share of Gender Specific Expenditure

Table 6.2 presents the federal expenditure on education under the recurrent and development budget for 2007–2008 and 2008–2009. In the recurrent budget the two major components that together constitute the largest part of the budget are government spending on higher education followed by spending on primary and secondary education, largely for Islamabad. The capital costs of building primary and secondary schools and cadet colleges are the major components of the development budget.

Table 6.2 Gender targeted education expenditure, Federal Budget (Rs. in million)

	2007–08		2008–09
	BE	RE	BE
Education affairs and services			
Current	24,147	24,280	24,622
Development	24,509	21,283	24,270
Total	48,656	45,563	48,892
Targeted gender-based expenditures			
Women and girls	1,279	1,301	1,751
Men and boys	2,621	2,254	2,920
Targeted expenditures	3,900	3,555	4,671
Targeted expenditures as %age of education expenditures			
Women and girls (%)	2.6	2.9	3.6
Men and boys (%)	5.4	4.9	6.0
Targeted expenditures (%)	8.0	7.8	9.6

Source Government of Pakistan (2009: 12ff)

Note BE refers to 'Budgeted Expenditure'; RE refers to 'Realized Expenditure'

Federal expenditure on education was greater for boys and men than spending on girls and women. Men-specific expenditure in 2007–2008 was 5.4 % of the education budget, while women-specific expenditure was just 2.6 %. In 2008–2009, the budget increased to 6 % for men and boys; and 3.6 % for women and girls.

Key initiatives in education

In order to promote education in general, and girls' education in particular, various initiatives were taken by the government, including the Education Sector Reform (ESF) Programme, the National Education Assessment System (NEAS), the Science Education Project and the establishment and operation of Basic Education Community Schools in the country, establishment of Cadet Colleges, Basic Education Community Schools and the introduction of M.Ed. classes at the Federal College of Education in Islamabad (Table 6.3).

The GBS examples demonstrate a large variety of approaches. The GBS can be an important document for highlighting gender and women's rights issues related to budgets and thus provide input for discussions on gender impacts of the budget in parliament or council, but equally can be an important source of information for civil society and thus promote transparency. Experiences gained from breaks in continuity of GBSs (e.g. Australia) show that it is important to produce a document that is accessible to a larger audience and contains meaningful information for policy makers and community at large. If governments produce the GBS, they might use it to boast about their achievements and merits and pay less attention to areas and issues where little or no progress has been achieved. Thus, the involvement of actors outside public administration is crucial. For example, the parliament could set standards and requirements for GBS formats and content. CSOs and/or researchers can cooperate with parliamentarians sensitive to gender issues by preparing policy briefs highlighting issues of importance to gender

Table 6.3 Pakistan's Education Sector Reform

Name of programme	Education sector reform (ESR) programme (2006–2010)
Overall objective	The main aim of the project is to enhance primary and elementary gross and net enrolment particularly for girls in Pakistan
Project components	…
Gender challenge	Poverty is both a cause and effect of the low level of education in Pakistan. The capacity and opportunities to earn higher income remain weak due to low education levels in terms of quantity and quality. Correspondingly low levels of income and poverty also constrain households/individuals from investing in the education for girls. As a result the gap has widened between the enrolment of boys and girls, while the achievement of the MDGs of universal primary education, the empowerment of women and gender parity is becoming harder to accomplish. The National Education Policy emphasizes increased enrolment in public sector schools, the removal of urban-rural and gender imbalances and the improvement of the quality of education at all levels. It focuses in particular on the implementation of the relevant curriculum reforms, strengthening of higher education, the provision of demand-driven education and encouraging private sector participation
Planned activities	…
Budgetary allocation	…

equality. Also, civil society can produce shadow GBS on its own, as is done in Austria and Canada, where gender aspects are included in broader alternative budget statements prepared by civil society coalitions, demanding broader transformations of budgetary policies.

For example, in South Africa, the Women's Budget Initiative cooperated closely with parliamentarians and provided them with GRB analysis-based support material in the form of a Women's Budget, at the time of budget deliberations in the parliament. This helped empower the parliamentarians in highlighting gender inequality and pursuing (or lobbying for) adequate resource allocations for issues related to gender equality and women.

In Uganda, CSOs, especially FOWODE, worked in continued exchange with parliamentarians to support and empower them to take an active part in budget discussions, trained them in GRB and provided them with policy and budget analysis. In Uganda, the Budget Office (BO) plays an important role in providing parliamentarians analysis and information on budgets from a gender perspective. The BO is particularly important in ensuring sustainability as it is a permanent institution located in the parliament and thus has regular staff and finance and is involved in the budgeting process according to Ugandan budget law. This is a major advantage for promoting GRB on a regular basis.

The UK Women's Budget Group (WBG) presents regular gender analyses of the draft budget and main policy proposals, with the aim of influencing budget deliberations. CSOs across the world can produce shadow or alternative budget reports at the time of the presentation of the draft budget in the parliament. These can be specific gender or women's budgets (e.g. South Africa) or a strong gender perspective integrated in a broader alternative budget (e.g. the Alternative Budget in Canada or the Civil Society Budget for the Future in Austria).

CMFEA, a Brazilian CSO, closely follows budget allocation to expenditure that is important to women (e.g. preventing and combating violence against women). After the draft budget is presented, the CSO lobbies members of parliament to push for increases in budget allocations for specific items. They are often successful, as Table 6.7 (titled 'Monitoring of budget allocations by CMFEA, Brazil') shows that the amount approved by the legislature is consistently higher than the amount in the draft budget law presented by the government for the period observed in this example.

6.5 Integrating GRB and Participatory Elements at the Budget Execution and Policy Implementation Stage

The budget execution stage involves implementing policies and approved budgets during the budget year. Implementation offers ample room for GRB work. At this stage, it is crucial to understand how budgets and policies are implemented, e.g., the manner in which services are delivered and investments are made. The main activities typically carried out during the fiscal year of budget and policy implementation are related to implementation of policies/programmes by line ministries and government agencies, delivery of services and transfers, investments undertaken and respective expenditures as well as implementation monitored. Furthermore, public revenues, e.g. taxes and user fees, are collected. This stage too offers potential for the integration of GRB. The main possible entry points at this stage are:

- gender responsive implementation of policies, programmes and activities;
- invited participation of women and men in the community to shape service delivery according to their needs respectively, community mobilizations to engage in participative activities;
- gender sensitive performance objective/indicators;
- collection of sex-disaggregated data and data on specific gender issues (e.g. violence against women, maternal health, gender pay gap);
- GRB analysis as a basis for the implementation of policies;
- involvement of beneficiaries in improving the quality of service delivery; and
- reporting that includes gender indicators and information on the impact of policies/programmes on gender equality.

A crucial question with regard to policy and budget implementation is whether citizens are merely seen as beneficiaries or as rights holders. The key actors in the budget and policy implementation stage are line ministries, sectoral departments and government agencies. Public officials responsible for the delivery of different public services, investments and other activities have a significant role to play. The private sector also contributes when the delivery of certain tasks is contracted out. The Finance Ministry or department has the vital role of releasing appropriate funds and, depending on the country's specific regulations, to approve larger projects.

At this stage, civil society engagement can be in the context of monitoring service delivery and demanding accountability. Civil society or researchers can also initiate GRB analysis of select government programmes to provide input for improved implementation or mobilize around key gender equality issues of importance to the community. To implement GRB by public administration in a meaningful way, it is critical to incorporate the perspective of potential and actual recipients of government services and allow for participatory processes in deciding about the implementation of public services, benefit and subsidy schemes as well as investments.

6.5.1 Examples of GRB Work at the Budget Implementation Stage

At the budget implementation stage, a broad range of activities can be undertaken to make policies and budgets gender responsive. These activities aim for public policies, programmes and activities to be implemented in a gender responsive way and the allocated budget to be used to promote gender equality and women's rights.

GRB analysis is a prerequisite at this stage to improve public policies and make them more gender responsive. A wide range of GRB instruments is used at this stage to analyse whether public services, public benefits and subsidies, public investments and other public activities are implemented such that gender equality and women's rights are promoted.

Equally, a wide range of participatory approaches exist in the context of GRB to enhance substantive gender responsive participation in policy and budget implementation processes. In view of the many possibilities which might be more well-known, this section is limited to one example in the context of gender responsive participation. At the budget preparation and implementation stages, participatory activities can be undertaken to involve women and men equally in implementation, e.g. in decision making on investment priorities. An example of gender responsive participatory processes is the work conducted in the municipality of Elbasan in Albania. The objective was to secure a more inclusive participatory process for women in the frame of a regular participatory budgeting process. To ensure that women's voices and needs are heard and dealt with, an assessment of the ongoing participatory budgeting process was undertaken with the aim to determine whether and how women were involved. Subsequently, a strategy was developed to address the imbalance between women and men in participation and to increase women's quantitative and qualitative participation in the participatory budgeting process. The aim was not only equal participation of women and men in numbers, but especially giving equal value to women's priorities and thus aimed at changing relations of power and influence. This involved practical aspects like gender aware information campaigns, working with women to encourage their participation, holding meetings at times and places that ensured that both women and men were able to participate and using gender sensitive techniques of moderation of town hall meetings and evaluation of preferences as well as changes in decision-making procedures.

An example of a useful method of analysis of selected government programmes to develop recommendations to improve implementation can be taken from a pilot project in the Autonomous Province of Vojvodina in Serbia.

6.5.2 GRB Analysis of Active Labour Market Policies in the Autonomous Province of Vojvodina, Serbia

This example presents a gender budget analysis of training services for unemployed people and entrepreneurs starting a business in the Autonomous Province of Vojvodina (Provincial Secretariat 2009; UNIFEM 2010). The objective of the programme was to support and stimulate the development of centres training the unemployed to return to the labour market and entrepreneurs to start their own business, and making them more competitive. Programme activities comprised training in computer skills (basic and several specialized computer skills) and training for beginner entrepreneurs.

Table 6.4 Participants according to the type of training and by sex

Training type	Training sessions	Participants	Women	Share of women (%)	Men	Share of men (%)
For the unemployed	48	504	320	63	184	37
1. Basic computer skills	19	240	198	82	42	18
2. Specialized computer training	29	264	122	46	142	54
For budding entrepreneurs	11	255	107	42	148	58

Source Provincial Secretariat (2009)

Initially, a situation analysis of women and men in the labour market was conducted along with a review of the criteria for participating in the programme. The situation analysis showed considerable gender differences in patterns of unemployment and entrepreneurship; the unemployment rate was 22 % for women and 15 % for men, while women constituted the majority of long-term unemployed persons. Only 23 % of the entrepreneurs were women while 77 % were men (all data for 2006). One element of the analysis of training services was the number of women and men participating in different types of trainings. Thus, availability of sex-disaggregated data on the users of services was important for the analysis. The number of participants, disaggregated by sex, in the different types of training is shown in Table 6.4.

The other important information is the cost for providing these services. This case refers to the training costs per participant. This was calculated based on information on the overall cost of the different types of training, including the trainers, facilities, material, etc. The average unit cost of the different types of training was arrived at by dividing the overall expenditure for a particular course by

Table 6.5 Distribution of resources—cost of training by type and average cost per participant (women and men)

Training type	Cost per participant (in RSD)	Women	Men	RSD spent on women (in 1,000s)	RSD spent on men (in 1,000s)
Basic computer skills	12,575	198	42	2,490	528
Specialized computer training: Web design	40,714	42	48	1,710	1,954
Specialized computer training: AutoCAD	48,060	22	46	1,057	2,210
Budding entrepreneurs: managing and improving business	21,210	100	140	2,120	2,970
Total		362	276	7,377	7,662
Average expenditure per woman/man				20	28

Source Provincial Secretariat (2009)

Table 6.6 Results—employment for participants after completion of training

Training type	Basic training (women)	Basic training (men)	Special training (women)	Special training (men)
Total number of participants	198	42	122	142
Training completed	198	41	120	141
Employment after training	67	13	67	63
	34 %	32 %	56 %	45 %

Source Provincial Secretariat (2009)

the number of participants. It is also possible to determine the expenditure on women and men participating in the training. As Table 6.5 shows, unit cost varied according to the type of training. The cost of training per participant was lowest for basic computer training and highest for specialized computer training in AutoCAD. Based on the number of women and men participating and the unit cost of training, public expenditure can be calculated by multiplying the unit cost by the number of women and men. The results for selected types of training are shown in Table 6.5.

Based on the results, a review can be conducted to ascertain whether the allocated funds have been spent efficiently in increasing gender equality. In this case, the analysis highlighted a few points that required further inquiry. Women evidently participated to a greater degree in basic computer training which was less expensive. Men's participation was higher in more expensive training courses. It is also important to combine these findings with the contribution made by training participation in the likelihood of finding a job in the labour market. Table 6.6 presents results on the likelihood of employment after the training. Not surprisingly, this is much higher for participants with a specialized training. Data shows that women participating in special training are more likely to get a job (56 %) than men (45 %). These results require further analysis of the barriers impeding women from participating in specialized training. There is a need for a better use of available resources.

Quantitative analysis was complemented by participatory approaches involving service beneficiaries and providers to learn more about the perspectives of women in training and the specific problems they faced in the labour market. Focus groups were organized separately with the public staff managing the programmes, participants, entrepreneurs and unemployed people. They helped gather qualitative data about perceptions, evaluations and specific needs or problems faced by women. The findings of this particular focus group highlighted the unequal initial conditions for entrepreneurs (lack of property) and the lack of support within the family, especially the lack of systematic support for women. Public officials managing the programme pointed out that the lack of a comprehensive survey of labour market conditions was an obstacle for programme planning and for the entrepreneurs themselves.

Based on the analysis and recommendations for changes in specific employment-related issues, programmes were developed with a view to making them more gender responsive. Such analyses are important and the involvement of beneficiaries of public services and resources support a more gender responsive implementation of budgets and policies.

6.6 Integrating GRB and Participatory Elements at the Budget Audit and Evaluation Stage

Finally, the audit and evaluation of the budget, carried out after the budget year has ended, usually involves independent auditing institutions and/or other auditing and evaluation processes within government. The main activities typically carried out at this stage of the budgeting process are:

- audit of revenue and expenditure (fiscal audit);
- formal auditing and reporting to the Parliament or Council;
- evaluation of the performance results (achievement of outcomes); and
- formal performance reporting to the Parliament or Council (annual reports).

While in the first years of GRB implementation, the stage of auditing and monitoring has been widely neglected in GRB work, this has changed, especially in the context of performance and results-oriented budgeting. The main possible entry points for GRB activities at the time of auditing and evaluation of budgets and policy implementation are:

- audit of spending on issues concerning gender equality and women;
- gender sensitive audit mechanism;
- gender sensitive evaluation criteria;
- performance reporting including information on gender equality performance;
- participation of citizens and gender advocates in evaluation and monitoring processes to include their views and experiences about budget and policy implementation;
- civil society activities outside the official procedures to mobilize around independent scrutiny of public expenditure and revenue auditing from a gender perspective as well as around independent evaluation of results (not) achieved; and
- debates on the achievement of gender equality when the audit report is presented to parliament.

Key players at the auditing and evaluation stage are auditing agencies, the government and line ministries or departments that conduct (internal and external) audits and evaluations as well as the parliament or council, which receives and debates the respective reports. Civil society can play an active role at this stage through participating in monitoring activities and/or organizing independent audits and monitoring to demand accountability.

6.6.1 Examples of GRB Work in Budget Audit and Evaluation

In Vienna, Austria, a gender budget audit document is presented annually in the context of the regular budget audit. It follows the format of the GBS and presents results of policy implementation. An important example of GRB work at the audit and evaluation stage is the inclusion of gender equality performance analysis in the context of regular performance analysis.

At the Federal level in Austria, a system of performance reporting and evaluation, with particular focus on gender equality, involves different institutions, namely the Ministries, in internal performance evaluation, the Federal Chancellery as a coordinating institution of performance evaluation of all ministries which provides regular reports to the Parliament, the Court of Auditors and the Budget Office in Parliament which supports parliamentarians. This system helps to regularly monitor the achievements of outcome objectives, including gender equality objectives defined at the budget preparation stage. In terms of its effectiveness, the quite elaborate system depends currently on the quality of the performance objectives and indicators defined by the ministries itself, which are rather mixed in its level of ambitions.

A monitoring approach for a particular budget item of high relevance to women's rights is presented as a final example. In Brazil, a CSO, CMFEA, monitors budget allocations for items of particular importance to women as part of its activities to push government to give more attention to key issues of importance to women's rights along the whole budget cycle. It is an example of continued participation by irruption, whereby CMFEA has created spaces for its involvement along the budget process. Table 6.7 presents an example of budget allocations for the programme on preventing and combating violence against women. CMFEA monitors the allocations in the draft budget (first column) and, if these are considered insufficient, lobbies parliamentarians to increase budget allocations. Table 6.7 shows that lobbying activities have been successful and budget allocations were increased by the

Table 6.7 Monitoring of budget allocations by CMFEA, Brazil—resources planned and executed for a programme on preventing and combating violence against women (in 1,000 Brazilian Real)

Period	Draft budget law	Authorized by legislature	% increase in allocation (%)	Paid values	Executed (paid) percentage (%)
2004	7,200	10,528	46	5,690	54
2005	8,222	10,135	23	7,894	78
2006	5,675	14,115	149	6,483	46
2007	8,109	23,545	190	12,279	52
2008	28,500	28,833	1	16,909	56
2009	28,844	40,909	42		

Source CFEMEA (2009)

legislators from 2004 to 2009. In this case, action is taken in the budget execution stage as well. The column 'paid values' shows that CMFEA continues to monitor activities at the budget execution stage by monitoring how much money has actually been spent on the programme under consideration. According to Table 6.7, this is an important activity because the money spent is often less than the money allocated. CMFEA can use this information for advocacy and lobbying activities and it follows this approach for various programmes of importance for women and gender equality.

6.7 Conclusion

This chapter illustrates the wide range of opportunities available to link GRB to the regular budgeting process at a very practical level, namely integrating gender equality, women's rights and restructuring processes to allow for participation as well as the opportunities for civil society to mobilize and create spaces of participation around gender equality and women's rights issues in the budget process. Even though the approach here is somewhat pragmatic in the sense of pointing out how GRB can be linked to the regular GRB process in normal administrative work, it is still relevant whether the actual GRB work done carries seeds for transformative dynamics or whether it remains at a static 'add on gender and do business as usual' approach. Some basic necessary criteria of a transformative mainstreaming approach to GRB are that entry points at each stage of the budget process are actually used to ensure that all actors regularly involved in budgeting and planning fully integrate a gender and women's rights perspective in budget preparations, adoption, implementation and financial and performance auditing and evaluation. In addition, actual changes in budget allocations, policy implementation, planning and budgeting processes as well as enhanced participation in decision-making processes is required. In order to unfold its full potential and effective promotion of gender equality and women's rights throughout the whole budget process, macroeconomic perspectives as well as meso and micro level need to be within the scope of GRB work. This is an ambitious agenda to which no current GRB initiative inside government fully lives up. Actually, the shortcoming of all of the wide range of examples of government GB work used here as illustration is that none fulfils this, as they disregard the macroeconomic perspective. As long as the implications of macro-economic policies and structural reforms are not integrated in GRB work, it will be piecemeal and not live up to transformative ambitions.

Mainstreaming gender is beneficial but is also a demanding and complex process of political change. Public policies beyond the scope of social policy, especially budgeting, are normally undertaken in a gender blind manner. Thus, adopting a gender responsive approach is a huge challenge for public officials. It demands a clear expression of political will and commitment by political leaders and top public officials to promote gender equality and women's rights. However, this is not sufficient as a change in perspective is necessary, as are new modalities of cooperation across departments and institutions, to ensure close synchronization among budget, gender and policy experts. Furthermore, a profound implementation of GRB will not be possible without opening budget processes for gender responsive participation at all stages.

As such, mainstreaming GRB is a broad change process that needs strategic planning and guidance. To facilitate this, it might be helpful to create a coordination mechanism, e.g., a GRB coordination group, involving top officials from strategic planning, budgeting and gender institutions as well as other key actors, e.g. from policy departments, CSOs and the parliament.

Developing a focused capacity-building strategy, involving training and practical support for all involved public officials and other actors from the very beginning, is also crucial. A limited number of entry points can be selected to make the initial GRB process manageable. This might involve ensuring that GRB is included in the budget call circular and the elaboration of a budget statement based on selected GRB analysis. A strategic perspective must be developed from the beginning to fully implement GRB at all stages of the budget process over a specific period of time.

GRB involves a longer-term strategy of change and is most promising when it is firmly integrated throughout the regular planning and budgeting process involving regular actors as well as gender advocates and civil society. Only then can a sustainable change process, aimed at achieving gender equality and women's rights and more effective public policies and budgets, as well as gender responsive participation be put in motion.

References

Budlender, Debbie, 2007: "Gender-Responsive Call Circulars and Gender Budget Statements", Guidance Sheet Series, No. 1, January.

Budlender, Debbie; Sharp, Rhonda, 1998: *How to Do a Gender-Sensitive Budget Analysis: Contemporary Research and Analysis* (Canberra-London: AusAID Commonwealth Secretariat).

CFEMEA, 2009: "Presentation of Brazilian Experience on GRB Work by CFEMEA", Powerpoint (New York: CFEMEA).

Elson, Diane; Cagatay, Nilufer, 2000: "The Social Content of Macroeconomic Policies", in: *World Development*, 28,7: 1347–1364.

Elson, Diane, 1999: *Gender Budget Initiative*. Background Papers (London: Commonwealth Secretariat).

Elson, Diane, 2006: *Budgeting for Women's Rights. Monitoring Government Budgets for Compliance with CEDAW* (New York: UNIFEM).

Government of Pakistan, 2009: *Gender Budget Statement 2008–09* (Islamabad: Ministry of Finance).

Hadžiahmetović, Azra; Đuric-Kuzmanović, Tatjana; Klatzer, Elisabeth; Risteska, Maria, 2013: *Gender Responsive Budgeting* (Sarajevo: Textbook for Universities, University Press).

Mahbub, Nadeem; Budlender, Debbie, 2007: *Gender Responsive Budgeting in Pakistan: Experience and Lessons Learned* (Islamabad: UNDP—Pakistan Government).

Ministry of Finance, Planning and Economic Development, 2013: *Budget Call Circular on Preparation of the Budget Framework Papers and Preliminary Budget Estimates for the FY 2014/15*, 27 September (Kampala: Government of Uganda).

Provincial Secretariat, 2009: *Towards Gender Budgeting in the Autonomous Province of Vojvodina* (Novi Sad: Provincial Secretariat for Labour, Employment and Gender Equality).

Sharp, Rhonda, 2003: *Budgeting for Equity: Gender Budget Initiatives within a Framework of Performance Oriented Budgeting* (New York: UNIFEM).

Steger, Gerhard, 2011: "Gender Budgeting. The Austrian Federal Experience", Presentation, Dar es Salaam, 21 June.

UNIFEM, 2007: *Gender Responsive Budgeting in FYR Macedonia.* Fact sheet, September 2007 (New York: UNIFEM).

UNIFEM, 2010: *Gender Responsive Budgeting in South Eastern Europe. UNIFEM Experiences* (New York: UNIFEM).

Chapter 7
Malaybalay City Integrated Survey System: A Tool for Gender Responsive Budgeting in Local Governance

Herculano S. Ronolo

Abstract Malaybalay city in the Philippines piloted a data-based system of local governance that is also useful as a tool for gender responsive budgeting. By collecting sex-disaggregated data about household membership, nutrition levels, education, income and other parameters of poverty, the system allows local government to identify gender issues and subsequently justify budgeting for social initiatives such as education, health and gender-sensitive livelihood training. The process of data gathering was also made gender sensitive and empowering by training the *barangay* health workers, many of them women, in collecting and processing the related information. Such an analysis allows us to ensure that budgets are not merely gender sensitive, but also accountable.

Keywords MISS · Sex-disaggregated data · Gender awareness · *Barangay* workers

7.1 Introduction

This chapter is based on the experiences of the city government of Malaybalay, Bukidnon, Philippines, in the use of a survey tool for improving its programme and service delivery. In particular, this chapter focuses on how the Malaybalay Integrated Survey System (MISS) has been used as a tool for gender responsive budgeting and has made the local government unit more result oriented.

First, the chapter explains the basis of the survey tool, its questions, its enumerators, its coverage, its system and other details. It focuses on sex disaggregation and how it helps identify gender issues. Next, the process of gathering data and information is discussed, including why it is considered an empowering and

---

Herculano Ronolo, City Administrator, Malaybalay City, Philippines. Email: allanronolo_56@yahoo.com.

C. Ng (ed.), *Gender Responsive and Participatory Budgeting*,
SpringerBriefs in Environment, Security, Development and Peace 22,
DOI 10.1007/978-3-319-24496-9_7

competency-building process along with issues and concerns arising from its use. The process flow of gender and development budgeting of the local government unit using MISS data is then highlighted. Finally, lessons learned from MISS implementation are discussed, including recommendations on measures to ensure its sustainability.

7.2 What Is MISS?

The Philippine national government developed the Community-Based Monitoring System (CBMS) in the early 1990s to improve governance and promote greater transparency and accountability in its resource allocation. The CBMS is a survey tool designed to collect information that policy makers could use to keep track of the impact of various government programmes and policies, especially among the marginalized section of society. The CBMS can capture the multidimensional aspects of household poverty through 14 measurable indicators: maternal and child mortality, child malnourishment, lack of access to safe water and sanitary toilets, presence of informal settlers, makeshift housing, incomes below food and poverty threshold, food shortage, victimisation due to crime, unemployment, elementary and high school participation.

In 2007, the city government of Malaybalay, a local government unit, seriously considered adopting the CBMS. However, several departments of the city government—from City Planning and Development to Health, Social Welfare and Development—reviewed the CBMS and found that it did not address some of their concerns.

Thus, in the last quarter of 2007, the Malaybalay Integrated Survey System (MISS) was conceptualized. It builds on the information requirements of CBMS but has additional questions customized to address the needs of the different departments of the local government unit, which are the primary end-users of the survey results. The CBMS contains 135 questions, while the MISS contains 231 questions. For example, one important MISS indicator is the absence of birth certification which is a concern for PhilHealth enrolment. Thus, periodically identifying persons in the *barangay*[1] could facilitate local civil registry efforts in birth registration. Another is the social welfare and development office concerns including occupational skills; number of children (3–5 years) attending early childhood care centres or preschools; families without health insurance; victims of natural calamities, disasters or accidents; and recipients and types of livelihood and scholarship programmes. The health office uses data on infant immunization, de-worming, supplements, nutritional status (for 6–12 year olds as well), breastfeeding data and supplemental food for malnourished children. Effective maternal care sought information on family

[1]The smallest political unit in the city, it is governed by a *barangay* captain and seven *barangay* councillors.

planning service providers, number of pregnant women, number of pregnant and lactating women who have had pre-natal and post-partum check-ups, vitamin supplements, tetanus vaccines and number of women with less than two years' gap between pregnancies. Relevant data was also collected with respect to community and environmental health, population development, licensing, veterinary care, agriculture department, and peace and order.

The MISS survey tool aimed at 100 % enumeration of the city's household population. For easier and effective interaction between enumerators and respondents, the survey questionnaire was written in the local dialect. The enumerators were *Barangay* Health Workers (BHW) who were residents of the area and thus familiar with its geography and people. Thus, they were considered most effective in asking questions and eliciting answers.

In-house programmers also developed a computerized system for encoding the information from the survey and processing the results. The BHWs, mostly women, also encoded the survey forms. Thus, through the MISS process, they were empowered and acquired the requisite competencies and understanding to address their community's problems.

The MISS establishes the universe of its constituents. As reflected in Fig. 7.1, critical information collected includes population distribution per *barangay*, (A) sex disaggregation, (B) household population including (C) average number of people

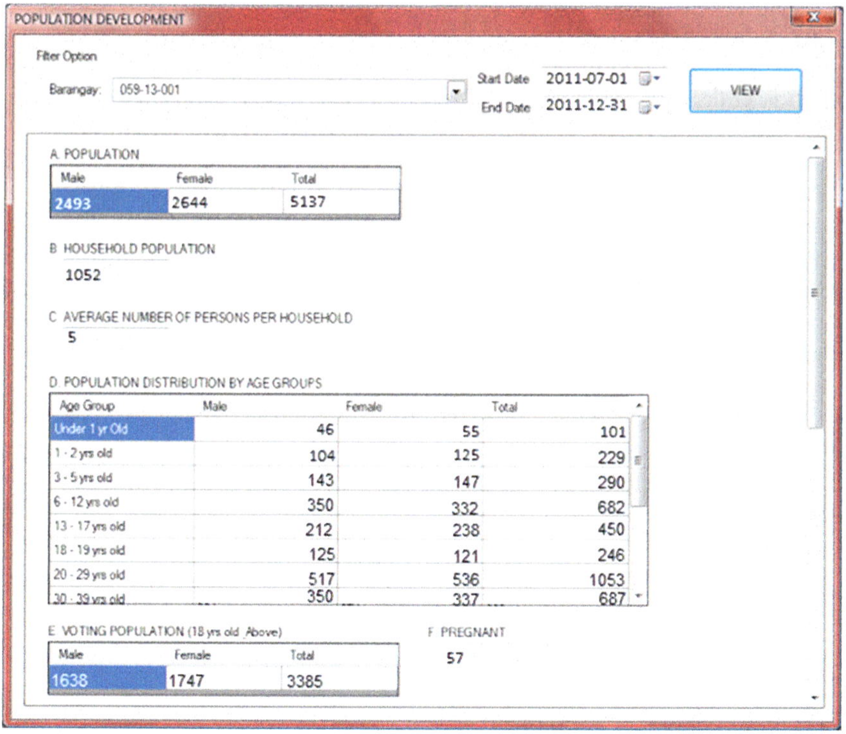

Fig. 7.1 Important demographic data collected. *Source* MISS, CPDO

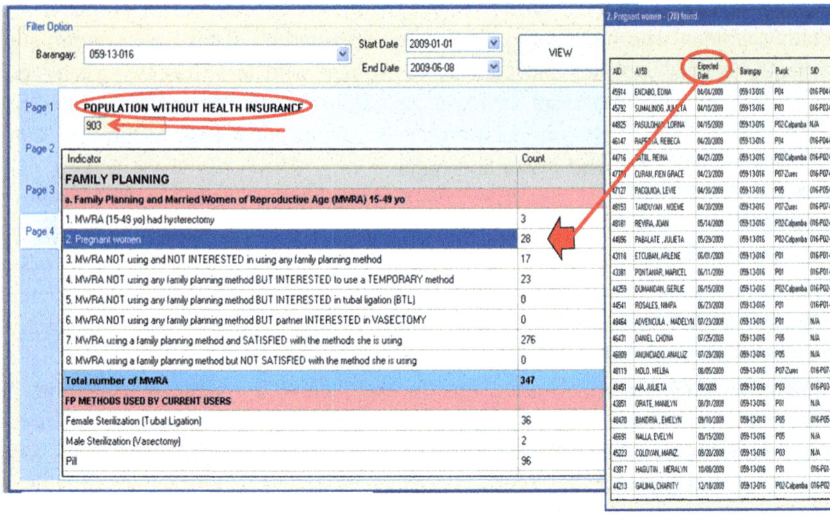

Fig. 7.2 Other important data captured by MISS. *Source* MISS, CPDO

per household, (D) population distribution by age including (E) voting population and (F) pregnant women for that period.

Apart from determining the number of pregnant women, the MISS could also process additional information such as the expected date of delivery (Fig. 7.2), which is critical for midwives and BHWs for timely intervention, such as mobilizing husbands to donate blood and offering screening services for newborns.

Individuals without health insurance can also be identified (refer to Fig. 7.2). This information is valuable not only to midwives and BHWs but also to the Local Government Units (LGU) for targeting beneficiaries of health programmes. Sex disaggregation is not restricted to population distribution but is applied to other indicators such as the number of unregistered children; school participation; victims of calamities, crimes, and accidents; occupational skills; recipients of livelihood and scholarship programmes; immunization data; nutritional status; tuberculosis patients; smokers and migrants, among others.

Such sex disaggregation gives the city government quantitative statistical information on the services provided to or received by women and men in the city. The resulting figures will determine if there are inequalities in the treatment of women and men. This then is the basis of analysis in identifying appropriate programmes and policies to overcome gender inequality (Figs. 7.3 and 7.4).

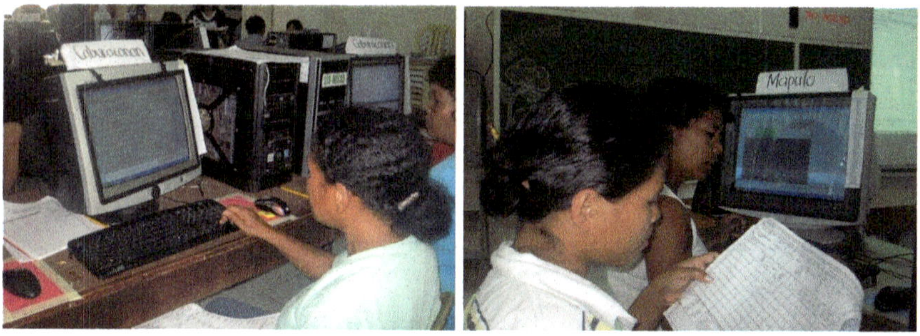

Fig. 7.3 The pictures show BHWs during the computer literacy training prior to their encoding task. *Source* CPDO

Fig. 7.4 MISS data in use. *Source* CPDO

7.3 MISS as an Empowering Tool

In 2007, as part of MISS implementation, the BHWs were tasked with conducting the survey, encoding completed forms and analysing the results. They were familiar with filling in forms and manual tabulation of data because of their work at the *barangay* health station. However, very few had experience in using computers. About 98 % of BHWs were women and many had not completed elementary education. Thus, they were initially reluctant to implement the MISS because of their lack of computer knowledge. However, all were willing to learn new things.

The software developed by in-house programmers was simple and used the same format as the survey intake forms, which the BHWs were already familiar with. Given the simplicity of the software, the critical requirement for BHWs to encode was computer literacy, especially use of the keyboard and mouse for encoding.

Thus, the first activity after the survey was to conduct computer literacy training for the BHWs. After the BHWs encoded the forms, in-house programmers with the city government helped them print the results. Computer literacy training, encoding and the MISS process helped build the BHWs' capabilities and confidence. In a meeting, BHWs, mostly women, noted:

> I am very happy that I know now how to use the computer. I feel that we (BHWs) are given importance because the city government spent for our training outside our *barangay*, for us to learn how to use the computer and provided us with a beautiful venue and great food (BHW from *Barangay* Busdi).
>
> I feel that our role is very important in planning for development in our *barangay* because the results of the survey we conducted and helped encode and process are used as the basis for planning programmes and projects for our *barangay* (BHW from *Barangay* Casisang).
>
> I am happy to help the city government in its effort to improve the services to our community. However, I also feel apprehensive that the *barangay* and city government might expect more from us BHWs and we might fail because we are new to this kind of work (BHW from Poblacion 1).

The entire MISS process including conceptualization, implementation, evaluation and participatory planning involved critical stakeholders, rendering it more sustainable because it encouraged ownership. In addition, the city government purchased 46 desktop computers that contained the MISS database and deployed them to all 46 *barangays* of the city for updates and maintenance. The deployment and technology transfer of MISS to the *barangay* is expected to contribute to its sustainability because it will enhance the skills and competencies of and transfer of ownership to BHWs and *barangay* officials who are the primary users of the information generated.

7.4 MISS: Collaborative Partnership

The city planning and development office (CPDO) spearheads MISS implementation. However, the actual work—from conceptualizing to conducting the survey, processing results and planning interventions—involves stakeholders such as the different concerned departments, *barangay* officials and BHWs. The activities conducted as part of MISS are discussed below.

7.4.1 Ascertaining Departmental Information Needs

The very first activity conducted for MISS was eliciting departmental information needs to determine whether they had already been captured in the CBMS. This process involved city government departments that provide frontline services, such as city planning and development; health, social welfare and development;

agriculture; veterinary services; civil registration; education; population development and the mayor's office. They helped incorporate into the MISS additional questions that would address matters that had not been captured by the CBMS. They also helped translate the questions into the local dialect because the CBMS is in Tagalog, the national language.

7.4.2 Enumeration Training for BHWs

The next activity was training BHWs for enumeration. The BHWs were familiarized with the questionnaire and trained to conduct the survey. The training involved an enumeration simulation to test the questionnaires. Based on the simulation experience, some questions were paraphrased or changed to make them easier and more understandable in the actual survey.

7.4.3 Data Gathering and Development of Computer Software

Once the questionnaires were finalized, the actual survey was conducted by the BHWs and *barangay* officials. They coordinated with leaders of the different *purok* or *sitios*[2] to schedule the survey in their areas. The leaders also provided transportation for the enumerator BHWs to and from the location of the survey. Some *barangays* with bigger budgets even provided meals and snacks to their BHWs.

When the survey was conducted, in-house programmers simultaneously developed computer software to capture and process the information gleaned. It was designed to match the questionnaire format to make it easier for the encoders to transfer information from the survey form into the system. The departments involved also made tabular formats of reports they wanted to extract from the MISS which were important for processing the results.

Conducting MISS requires a substantial amount of money since it is a 100 % survey of households. The BHWs are paid for their work, computers have to be purchased and training has to be conducted. In addition, the survey also takes time. With 32,000 households in the city, a minimum of six months is needed for the MISS: around three months to complete the enumeration and another three months for encoding and processing results. Thus, it is only conducted once every three years. Ideally, it is scheduled in the year prior to the mayoral election so that when the new administration assumes office, it can be provided updated city information.

---

[2]*Purok* or *sitio* is a small village made up of 10–100 households and is part of a *barangay*. A *barangay* could have several *puroks* or *sitios*.

Table 7.1 GAD budget and expenditure

City of Malaybalay Gender and development budget and expenditures As of 31 December 2012			
Function/programme/project	Appropriation	Expenditures	Balances
Gender and development			
A. Gender Mainstreaming			
(a) Advocacy on RA 9262 Violence Against Women	100,000.00	100,000.00	0.00
(b) Reproductive Health and Motherhood	100,000.00	100,000.00	0.00
(c) Training and Seminars	100,000.00	39,850.00	60,150.00
B. Skills Training and Development Programme			
(a) RIC Skills Training Seminar	200,000.00	147,243.86	52,756.14
(b) 4-H Club Capability Development Programmes	100,000.00	99,307.00	693.00
C. Emergency Funds for Medicine and Hospitalization. Hospital and Laboratory Fees for Indigents	8,575,994.06	8,555,129.03	20,865.03
D. Comprehensive Nutrition Programme			
(a) Micro Nutrient Supplement	200,000.00	199,950.00	50.00
(b) Deworming Programme	100,000.00	100,000.00	0.00
(c) Supplemental Feeding Programme	400,000.00	391,334.00	8,666.00
(d) City Nutrition Programme	1,000,000.00	937,688.21	62,311.79
E. Comprehensive Health Programme			
(a) BHW/BNS/Hilots Day	200,000.00	196,635.00	3,365.00
(b) CBMIS	300,000.00	283,411.00	16,589.00
(c) Support to Health Services PPAs	500,000.00	175,856.97	324,143.03
(d) Health Insurance of *Barangay* Officials	1,000,000.00	993,600.00	6,400.00
(e) Health Sanitation and Maintenance Programme	1,000,000.00	1,000,000.00	0.00
(f) Support to POPDEV	500,000.00	487,013.64	12,986.36
(g) City Indigent Programme	1,000,000.00	1,000,000.00	0.00
F. Comprehensive Education Programme			
(a) City Scholarship Programme	4,000,000.00	4,000,000.00	0.00
(b) Vocational Technical Skills Training	300,000.00	298,130.00	1,870.00
(c) Alternative Education and Functional Literacy	1,000,000.00	623,031.18	376,968.82
d) Support to Education	3,083,889.79	3,083,889.79	0.00
(e) SP Educational Fund	2,006,514.33	1,936,423.71	70,090.62
G. Support to Children and Women Organization			
(a) Child-Friendly Movement Programme	1,000,000.00	999,786.72	213.28
(b) Implementation of R.A. 8971 Solo Parent Welfare	100,000.00	9,000.00	91,000.00

(continued)

Table 7.1 (continued)

City of Malaybalay Gender and development budget and expenditures As of 31 December 2012			
(c) Support to Women's Day	200,000.00	200,000.00	0.00
H. City Council for the Protection of Children			
(a) Support to Committee for the Welfare of Children	100,000.00	15,250.00	84,750.00
I. Other GAD Activities			
(a) Financial Assistance to Senior Citizens	6,000,000.00	6,000,000.00	0.00
(b) Support to Persons w/Disabilities Programme	100,000.00	81,750.00	18,250.00
(c) Support to Clean Up Day	20,000.00	20,000.00	0.00
(d) Support to Barangay Health Workers	2,400,000.00	2,400,000.00	0.00
(e) Support to Day Care Workers	3,200,000.00	3,113,000.00	87,000.00
Total	38,886,398.18	37,587,280.11	1,299,118.07

Source Annual Investment Plan, Malaybalay City (2012)

7.4.4 Computer Literacy Training and Encoding

As mentioned earlier, the BHWs also encode the survey forms. This is to minimize data entry errors because it is assumed that the BHWs are familiar with the data. However, most BHWs only have elementary education, and computer literacy training helped the BHWs assigned to encode the survey results. The MISS process also encouraged cost sharing between the city government and the *barangay* units. This process began with conducting the survey and continued through computer literacy training and encoding. The city government shoulders the accommodation and food needs of the participants, while the *barangays* pay for their transportation.

7.4.5 Processing Results

After the encoding, in-house programmers assist the BHWs in processing the survey results and preparing them for presentation to the *barangay* council. The BHWs are the first to see and analyse the results with the help of the CPDO staff. They are also the first to learn of their community's problems as elicited by the survey and probably the first to think of possible solutions. However, *barangay* officials tend to blame the BHWs for unpleasant reports, attributing the unpleasantness to carelessly conducted surveys or BHWs' mistakes during encoding. Such mistakes are likely but these are discussed openly and mistakes rectified.

7.4.6 Presenting and Validating Reports

The BHWs then present the results to the *barangay* officials, who will either confirm them or ask for further validation when they are unsure of the results. For instance, if a report identifies a certain household that did not receive any livelihood assistance but the *barangay* officials can certify that the household was given such assistance, the BHWs would need to validate the information before finalizing the report. In most cases though, the *barangay* officials normally confirm or accept the report.

7.4.7 MISS for Development Planning

The validated and finalized survey results are used by the city's planning and development office to facilitate development planning workshops at the *barangay* level. The results become the basis for the *barangay*'s development planning, especially in identifying target beneficiaries of programmes and projects. To build the community's capability to document and create a basic profile of their *barangay* based on the MISS information, the CPDO arranges a session on documentation and basic socio-economic profiling. Some of the information is useful for re-planning purposes, as noted below.

The survey also brought gender issues to the surface. For instance, in the agriculture department, ownership of properties is in the name of husbands whose wives cannot enter into formal agreements on these properties. Although the law (women in development and nation building or RA 7192) gives women equal contractual rights, tradition weighs so heavily, especially in the rural areas, that women there say "My husband will decide". Another example: agricultural services lack women's perspective, recognizing women's role only in planting (in rain-fed and irrigated areas). During the off-farm season women become invisible without any activities to cater to their needs, e.g. training in crafts, food preservation. The city agriculture office has traditionally focused on men-led activities (Honculada 2009: 93)

7.4.8 Dissemination of MISS Results to User Departments

The different departments of the city government that initially identified the information they needed from the survey receive consolidated reports from *barangays* at the end of the process. The CPDO consolidates results from different *barangays* and makes it available to other departments in tabular form with numbers. Names are not shared, to protect the identity of the individuals. These use it not only as baseline information but, more importantly, in planning and focused targeting of programme beneficiaries.

7.5 MISS Uses and Evolution

The quantitative data generated from MISS is considerably helpful to the *barangay* unit and city government in development planning. To date, the information generated has been used in:

- preparing profiles and *barangay* development plans for 46 *barangays* in the city;
- preparing the city's ecological profile, which is a valuable input for formulating comprehensive land-use plans and comprehensive development plans (CDP);
- gender responsive budgeting;
- data analysis for formulating local poverty-reduction action plans;
- disaster risk management; and
- monitoring the LGUs' accomplishment of millennium development goals.

Lack of gender perspective and a bias towards a visible legacy means most officials prioritize infrastructure projects over social services. However, with a GRB initiative, a people-focused outlook is seen. After all, what good is a house if the occupants are sickly (Honculada 2009)?

7.5.1 Integrating MISS into the GIS

To make MISS more useful and to draw out more qualitative data, the MISS is integrated into the Geographic Information System (GIS). This helps paint a more complete picture of the community. For example, the location of the housing structure of a household is plotted into the map and the attributes of that particular household can be viewed in the GIS. The occupants of that household, including its members, information generated from the MISS survey, can also be viewed in the GIS.

With the MISS and GIS integration, information can be analysed down to the *purok* level. Thus, the city government will know if its services have reached far-flung areas. With sex-disaggregated data, it can also determine whether women in such areas are able to avail of services, especially given women's double and triple work burden. Usually, distant *puroks* are not reached by services. Physically seeing the households on the map helps include them in the analysis—a personal touch that the social data of MISS might not be able to accomplish.

Map digitizing of the survey means that a click of the mouse reveals where pregnant women reside (the area is coded red), in which *barangay*, and how many, and with another click, the house each one lives in, indicating her socio-economic status. Among the survey highlights: problems in peace and order, problems in livestock dispersal (the survey revealed that some *barangay* employees owned three heads of livestock which rightfully belonged to farmer-beneficiaries so livestock bidding was frozen), and problems in perception about needed infrastructure (*barangay* captains wanted satellite markets and yet existing *barangay* markets revealed only 30 % occupancy) (Honculada 2009: 93).

Also included in the GIS is the picture of the structure, which is useful not only for social service delivery but also helps the engineering office monitor building permits issuance and standards compliance, while the assessor's and treasurer's offices can use it for taxation purposes.

7.5.2 *Different Departments' Use of MISS Information*

As mentioned above, different city departments are given a copy of the MISS results consolidated by the CPDO. Some examples of how different departments have used the information generated from the MISS are given below.

- The civil registrar's office obtains not only the number of children not registered at birth, but also their names and location. Even the reasons for non-registration can be gleaned from the survey. With all this information, the civil registrar's office can target the specific location and number of children for mobile registration. It could also focus information to be disseminated to community members to encourage them to register.
- The department of education collects data on the number of school-age boys and girls—how many are out of school, where they are located and why they are not enrolled in school. Given the information, the department can determine where back-to-school campaigns must be conducted before enrolment and also plan for additional infrastructure, teachers and learning materials in areas where enrolment is expected to increase based on MISS data.
- Useful information for the social welfare office includes the number of single parents, senior citizens and households without any health insurance, among others. Using this information, the office can plan how many new senior citizens' cards it needs to prepare and the number and location of senior citizens who need assistance; how many households, and which must be prioritized, to enrol in the Philippine Health Insurance Indigent programme; and how many and where single parents, many of whom are women, are located, to organize delivery of various social services.

7.5.3 *Tool for Advocacy and Accountability*

Survey questions concern the services/benefits received from the city government such as scholarships, seeds subsidy, etc. Thus, the MISS integrated into GIS promotes government accountability because it can determine if the government is serving those who deserve assistance and not those who are politically favoured.

It can also be an advocacy tool for the city government's services because if people are unaware that such services existed, then through the survey process they can be informed. It can also inform others of the presence of the people themselves, especially when MISS data is shared with other national line agencies, non-government and private organizations resulting in collaboration and partnership to extend assistance or delivery services to the community.

If the survey is conducted regularly, every few years for example, then the progress or impact of policies and programmes introduced can be monitored, including the development of the people or the changes in the quality of living.

7.5.4 MISS Facilitates Gender Responsive Budgeting

No endeavour of the local government unit can be implemented without budget allocation. Thus, perhaps the most critical policy tool of the government is the budget. Without adequate budget allocation, no programme or policy can be implemented. Given this reality, the local government must also recognize its responsibility to ensure that the budget is allocated to policies and programmes that will address the needs of its constituents. The government also has to ensure that the budget reaches target beneficiaries, including women, and determine whether the desired result is achieved. A significant change was noted in that:

> ... before the introduction of GRB, women received training simply to meet targets with no real intention for them to apply what they learned. Currently, however, they receive the support needed to sustain these learnings. Women are able to proceed on their own with some amount of monitoring (Honculada 2009: 94).

It can be further argued that the budget should reach women who need it. However, this does not mean separate budgets have to be created for women and men. Instead, the budget should be a tool to bring greater awareness of the issues faced by women and men not only at the city level but also at the *barangay* level. In addition, it is already national policy in the Philippines for all local government units and different government agencies to allocate 5 % of their budgets to address gender and development (GAD) concerns. The challenge then is identifying appropriate programmes, projects and activities to be funded from the mandated 5 %. In Malaybalay, the MISS data was useful in identifying programmes, projects and activities for GAD budget allocation.

The MISS data for Malaybalay indicated that women needed more health services, particularly in the area of reproductive and child care. Employment data showed more unemployed women than men. Most often they could not work because they were taking care of children or other family members. These women could possibly be more economically productive and participate in societal activities if they had more time and opportunity to do so. This would then entail providing them with appropriate reproductive-health services that would free them from short-gap pregnancies. Adequate child-care services would give them more time and opportunity to be more economically productive. This need not mean that men's needs could be ignored but that women of reproductive age tend to have greater need in terms of health services than their male counterparts.

In other areas of MISS too, women are given greater attention, especially in education and social welfare and rural development. Remedios Sarzuelo, or Remy, Malaybalay's assistant city agriculturist, noted that gender awareness is changing the personae and lives of women farmers. As always, more women participate in seminars and farmers training, even during off-season. So it is not because men are busy with farming, though tasks requiring a visit to the 'centro' or city are undertaken by men. Earlier, recipients' lists used to have only one column for the names of farmers, invariably male, although wives often came in their stead. Now, another column has been added to record the names of women participants. Women are now counted because they have a right to be included, not just to comply with the GAD requirements.

There is a change in rural women who participate in gender awareness raising and livelihood activities such as cut-flower propagation, vermiculture, and backyard gardening: they have confidence to face people, they are articulate, and are able to seek assistance from politicians in a straightforward fashion. "In the past, we would have to seek them out in the *barangays*, now they respond to a simple SMS and come," Remy says. A husband's approval for them to attend seminars becomes a formality for these rural women, who now know, and act on, their rights (Honculada 2009: 94).

7.5.5 Data Processing and Analysis

For Malaybalay, the MISS paved the way for gender responsive budgeting. The data processing and analysis stage of MISS provides the government with data on people who have not been reached by its services and their locations. Sex disaggregation of data provides the identity of these people as well. For example, data showed that more women were unemployed because they were tending to their young children at home and that more adolescent boys were not in high school because they were already gainfully employed. This information clearly presents gender issues to consider in gender responsive budgeting.

Given this information, the city government could provide home-based livelihoods for women or provide additional child-care services to allow them to be more economically productive. In terms of education, the city government could provide scholarships for adolescent boys so they continue their schooling.

Based on the MISS findings mentioned previously, some items in the GAD budget have been changed. The biggest changes are in the Comprehensive Health Programme allocation which increased from approximately 900,000 pesos to 8.9 million pesos. New programmes that were allocated under the health budget include: (a) Indigency Programme which enrolled poor families for health insurance; (b) Honorarium to *Barangay* Health Workers; (c) Health Sanitation and Maintenance; and (d) Support to the Population Development Office that takes care of some reproductive health concerns of women. Another significant change of 143 % in the budget is for the Comprehensive Nutrition Programme, which added one million pesos for its implementation after MISS results revealed malnourished children in the remote areas.

The budget for education has also increased from the original budget of 12 million to 13 million pesos. Particular programmes added under education include the giving of Honorariums to Day Care Centre Workers; and the Alternative Learning System which are mostly trainings for women who are already out of school. Also worth noting in the GAD budget before and after MISS is that the programmes to be funded are now more specific.

Improving women's access to basic social services such as health, education and nutrition is one of the ways of responding to practical gender needs. However, the City Government wanted to bring its response to gender issues at the second level, which is by empowering the women. As NCRFW (2006: 54) specified in the handbook on Gender and Development, empowering women can be by means of giving them the tools and capabilities to provide the services themselves and for other women. As reflected in the city's GAD budget, the increase in education budget is aimed at increasing women's capabilities. In addition, the item on Support to Children and Women Organization increased by 100 %. Specific programmes under this item will ensure that children and women are given venues to participate in community activities so they can be empowered to express themselves and also help others.

7.5.6 Planning

The next stage is planning. At this stage, departments, the *barangay* and the city government review and analyse existing programmes, projects and activities. This is to ensure that problems and concerns identified by the survey are addressed and appropriate interventions are proposed, especially those that will address gender issues.

7.5.7 Resource Allocation

At the resource allocation stage, the proposed programmes, projects and activities that will address GAD issues and concerns will be allocated from the mandated five per cent. Gender budgeting is not solely for the benefit of women but focuses on budget allocation that will have a positive impact on the lives of both women and men.

7.5.8 Monitoring Results and Impact

The result of the next MISS cycle will give the city government an idea of the impact its programmes and projects have had. More specifically, it will hopefully identify where and how budgeted and allocated money is spent and whom it reaches. It could improve the situation of the people and have a positive impact on gender issues.

7.5.9 Sample of GAD Budget

Presented above is the 2011 GAD budget that considered MISS results in proposing programmes and activities to be funded. Areas that received allocation were health, education and social welfare because these were areas where women's concerns were most obvious (Table 7.1).

7.6 Conclusion

Since the first MISS survey in 2009, the CPDO that spearheads its implementation has taken note of several issues and concerns that have arisen from this learning process. MISS is an evolving process; thus, the city government can improve the system and adapt it to the current situation and needs of the city. Since it has been developed locally, survey questions and systems can also be modified and enhanced to suit the information needs of the city government. Local development also means immediate local availability, unlike the national census which is processed at the national level and the results take several years to reach the community.

MISS stakeholders, especially BHWs and *barangay* officials, can claim ownership of the system not only because they conducted the survey and analysed the data but also because they can use the data to plan for their *barangay*. In addition, they are also responsible for maintaining and updating the database. Not only does this arrangement render the data highly accessible to its users, it also empowers the BHWs in the process.

However, deployment of the MISS system to the *barangays* also takes the maintenance and repair of software and hardware to the *barangay* level. This is extra challenging because the *barangays* are located far from each other and only three people are capable of troubleshooting the hardware and software deployed to the 46 *barangays*. Encoding and processing survey results at the *barangay* level is thus quite difficult to supervise and monitor. The city government is considering centralizing the encoding and processing and deploying only the cleaned and finalized database to the *barangays*.

To improve the MISS, the city government is looking at online tracking of changes made to the database. Processing and updating information online would be faster and would save resources for the city. However, this might not be realized in the near future since there are still about 30 (out of 46 *barangays* in the city) that do not have internet connections.

Concerns over data integrity and confidentiality are currently addressed by a data-sharing protocol. Under this protocol, tabular MISS reports, containing only numbers, are readily available to anyone. However, only registered BHWs with assigned passwords can actually open the files with the names of people with the numbers.

The integration of MISS into GIS created opportunities for a unique overlap of qualitative analysis of communities' situations on quantitative data. It also added geographical analysis of issues and concerns which is useful for the land-use planning of the city government. This MISS/GIS information is used extensively as a baseline by various departments of the city government and by other national agencies, including education, health, social welfare and civil registration, to plan and identify target-focused programmes.

A drawback to the widespread implementation of the MISS could be the substantial investment required. In Malaybalay, with almost 35,000 households and 231 variables/questions to be processed, the MISS requires a budget of 3 million pesos (USD 68,800). At least 20 full-time employees are needed for six months for all tasks, from re-training enumerators to data gathering, encoding and processing results.

In terms of gender mainstreaming, a thorough analysis of gender patterns or issues highlighted by the survey is still lacking. So far, only data relating to health, nutrition and social welfare programmes, which are directly related to women, are being noted and analysed. Greater analysis of underlying gender issues is possible by relating one indicator to another. For instance, data shows that more women are single parents as compared to men. Why is this so? To what particular age group do these women belong? Are they still in school? What is their educational attainment? What is the situation of the men in the same age group? Such an analysis can highlight invisible links and suggest possible solutions.

The link between data analysis, planning/targeting and GRB has already been established. However, the local government is yet to determine whether allocated budgets are being spent as intended. By ensuring that allocations are well spent, the city government could also expect positive outcomes from its intervention, including in relation to gender equality.[3] This is a challenge for the next MISS cycle.

References

Honculada, Jurgette, 2009: *Gender in Good Governance: Examples of Local Innovations in Gender Responsive and Results Oriented Budgeting* (Bangkok: UNIFEM).

NCRFW, 2006: *Moving Forward with GAD: A Handbook on Gender and Development for the Sanggunian Committee on Women and Family* (Manila, Philippines: UNICEF).

[3]As Honculada (2009: 99) noted: "The Malaybalay City LGU has scored many gains in the attempt to 'genderize', and ensure resultant impacts from, its yearly budgets. Staff, both executive and clerical, seek to translate gender concepts into their daily life and work. Civil society organizations, prodded on by women NGOs, earnestly engage in the planning and budgeting process".

Part III
Institutionalizing GRB: Towards Better Accountability and Good Governance

Youth from the urban community speaking about their experiences in the GRB project. The photo was provided by PWDC who granted permission to include it in this volume

Chapter 8
Localizing Gender and Participatory Budgeting: Challenges of Institutionalization in Penang, Malaysia

Aloyah A. Bakar, Patahiyah Binti Ismail and Maimunah Mohd Sharif

Abstract Institutionalization is the introduction of new practices for sustained change, and it is a complex and difficult process. This chapter analyses the readiness of the two local governments in Penang to institutionalize GRB within their respective organizational milieus. It points out that lobbying with and sensitizing policy makers as change agents in the early stages of its formulation is an important pre-condition of institutionalization. At the same time, local authorities need to create an enabling and supportive environment to make GRPB a reality in their respective bureaucratic contexts although competing priorities might affect actions and commitment. The chapter argues that both a participatory approach and a commitment towards gender integration into budget structures and processes are the way forward.

Keywords Institutionalization · Gender and participatory budgeting · Pilot project · Municipality · Local government

8.1 Introduction

In January 2012, the Penang Women's Development Corporation (PWDC) together with the two local councils in Penang, Malaysia, the Penang Island Municipal Council (MPPP) and the Seberang Perai Municipal Council (MPSP), started a

Aloyah A. Bakar, Chief Executive Officer, Penang Women's Development Corporation, Penang, Malaysia. Email: aloyah.bakar@pwdc.org.my.

Patahiyah Binti Ismail, Mayor, City Council of Penang Island, Penang, Malaysia. Email: Datukbandar@mbpp.gov.my.

Maimunah Mohd Sharif, President, Municipal Council of Seberang Perai, Penang, Malaysia. Email: ydp@mpsp.gov.my.

C. Ng (ed.), *Gender Responsive and Participatory Budgeting*,
SpringerBriefs in Environment, Security, Development and Peace 22,
DOI 10.1007/978-3-319-24496-9_8

three-year pilot Gender Responsive Budgeting project (2012–2014). Its long-term goal was for the Penang Local Government to integrate gender perspectives into its governance processes, particularly through the implementation of Gender Responsive Budgeting (GRB).[1]

The project was kick-started as a flagship programme of PWDC, a state-funded women's agency established in late 2011 to promote good governance and gender equality in Penang.[2] Its project partners, however, have existed since colonial times but were re-structured, in 1976, as municipal councils under the Local Government Act (LGA).[3] This is the third and last year of the pilot. Hence, it would be instructive to evaluate the progress thus far and identify the challenges in moving this young endeavour forward.

This chapter discusses the nature and extent of the institutionalization of GRB in the two local councils.[4] It specifically analyses the readiness of the two local governments in Penang to institutionalize GRB within their respective organizational milieus.

Institutionalization, particularly the introduction of new practices for sustained change, is a complex and difficult process. The first part of the chapter attempts to briefly engage with this term. This is followed by a discussion on lobbying with and sensitizing policy makers as change agents in the early stages of its formulation—an important pre-condition of institutionalization. The key roles played by local authorities in creating an enabling and supportive environment to make GRB a reality in their respective bureaucratic contexts are then examined. The chapter points out how competing priorities might affect actions and commitment on GRB, as political will alone might not achieve sustainable GRB institutionalization. The concluding comments argue that both a participatory approach and a commitment towards gender integration into budget structures and processes must be the way forward. This will ensure the mainstream is successfully transformed towards gender inclusivity and sustainability of people-centred governments, Penang included. Some recommendations are then put forward for the next phase of the project.

---

[1]In 2004, the United Nations Development Programme (UNDP) and the Ministry of Women, Family and Community Development (MWFCD) piloted GRB in five ministries in Malaysia. Following this, three Treasury Call Circulars duly encouraged other ministries and agencies to use GRB in their programmes and activities. However, the latest 2014 call circular for 2015 is more strongly worded, requiring various ministries and agencies to prepare their budgets using a gender analysis budget approach (*perlu menyediakan perancangan bajet menggunakan pendekatan analisis gender*); see: http://www.treasury.gov.my. Thus far, Penang is the only one to initiate GRB at the state level.

[2]Registered in November 2011, PWDC only started operating in January 2012.

[3]Malaysia is governed by a three-tiered federal system. The federal government is the highest authority followed by the state government and local government.

[4]We would like to thank Cecilia Ng and James Lochhead for their valuable inputs to this chapter.

8.2 Institutionalization as Sustained Change

Levy (1996: 1), citing Abercrombie, Hill/Turner (1988), defines institutionalization as a "process whereby social practices become sufficiently regular and continuous to be described as institutions". These "social practices that are regularly and continuously repeated are sanctioned and maintained by social norms, and have a major significance in the social structure". As she correctly points out, the term has two important concepts, "that of the room for manoeuvre which individuals and organisations have to generate change, and that of the notion of sustained change". Levy further stated that the idea of "sustained change" could cause conflict between the regular practices of organizations with their own set of interests and how they respond to change, which also reflects other power relations and interest dynamics and patterns. Levy then propounds a rather intricate web of institutionalization, with 13 elements as conditions towards such sustained change.

Other authors have similarly pointed out the need to define the different rules, norms and practices, formal and informal, in the institutionalization process. For example, Helmke/Levitsky (2004: 727) define informal institutions as "socially shared rules, usually unwritten, that are created, communicated and enforced outside of officially sanctioned channels" as opposed to formal institutions which are "rules and procedures, that are created, communicated and enforced through channels widely accepted as official".

How does organizational change occur? Kelleher/Rao (1999) provide a useful framework to understand such changes, particularly when introducing gender issues into organizations. They point out that four interrelated clusters of changes need to be made. The top two clusters are individual while the bottom two are systemic. The cluster on the right is of formal institutional rules and the one on the left is informal rules and cultural practices that maintain inequality in everyday practices. Each quadrant has its own challenges with connections between them. The formal is visible while the informal is less visible and at times invisible (Fig. 8.1).

Clearly, change must take place at many levels, must occur holistically across different quadrants or layers or spheres, and will take time. Furthermore, taking off from Levy (1996), Kelleher and Rao argue, quite succinctly, that one can move an organization towards transformation in a web of five spheres (Fig. 8.2). These are:

1. *Politics* Is there a women's constituency advocating gender issues that will be taken up by the organization?
2. *Organizational politics* Are these external gender advocates able to negotiate with those wielding power (i.e., senior bureaucrats) within the organization to adopt these issues? The outcome of bureaucratic buy-in could be a stronger policy or increased resources, including the possibility of an alternative organizational culture.
3. *Institutional culture* Organizations have an institutional culture with values, history and methods for doing things. This culture might facilitate or impede gender equality efforts. To what extent is there a culture of openness, dialogue

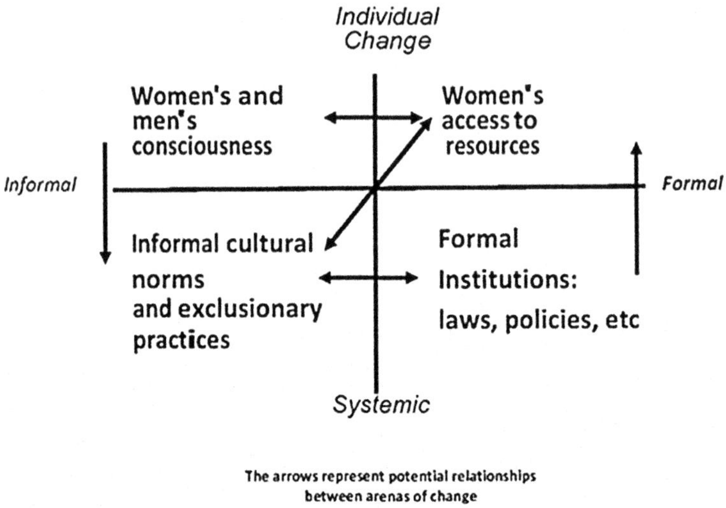

Fig. 8.1 What are we trying to change? *Source* Kelleher/Rao (1999: 2)

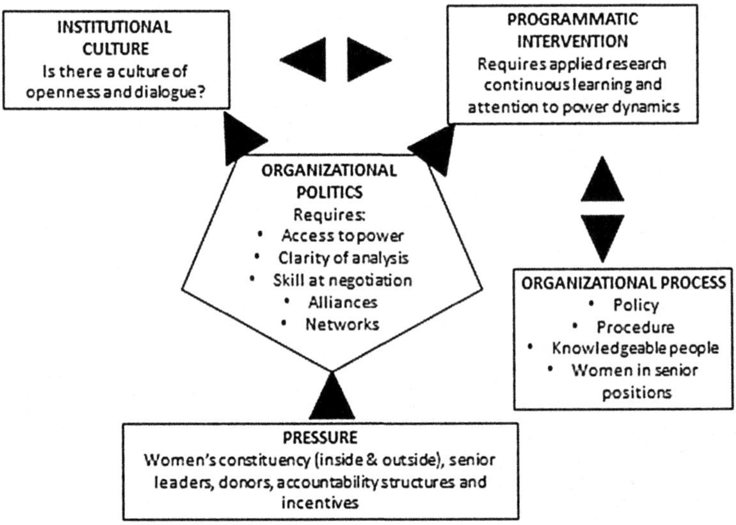

Fig. 8.2 Organizational likelihood of promoting gender equality. *Source* Kelleher/Rao (1999: 6)

and understanding for new directions, specifically with respect to gender issues and programmes such as GRB?

4. *Organizational process* Are sufficient resources and skilled and knowledgeable people available to lead the change, particularly in terms of learning new ideas and creating new programmes, policies, services and structures? For example, is

gender analysis a key component of all projects? How does this translate into work on the ground?

5. *Programmatic interventions* Have new methodologies/applied research been developed to ensure that gender equality efforts are appreciated and supported by those in the organization? Can these new methodologies make a difference? This is the test for whether the organization delivers value or not and whether divisions support each other in these efforts.

This chapter selects some of the above-mentioned concepts, particularly the framework of the five spheres of power/influence, to examine the challenges faced in institutionalizing a new strategy—that of GRB in Penang, Malaysia. Similarly, Illo et al. (2010: 7) provide some useful insights in their evaluation of the Gender and Development (GAD) Budget Policy in the Philippines. They point out how gender budgeting is a triple process: political, technical (under gender mainstreaming) and administrative (a part of broader governance change).

The next section starts by documenting the initial stages of the project and the steps taken to obtain buy-in (including securing funds) from the authorities. This can be categorized as the first and second sphere of influence where women advocates engaged in dialogues from mid-2010 with the state and Local Authorities (LAs) leadership before they agreed to adopt GRB as a pilot in late 2011.

Following this, the third part of the chapter discusses the relevant outputs of the pilot, highlighting the importance of community pilots and the innovative methodology employed for GRB acceptance in terms of 'bringing value' to both LAs, the last sphere of influence. It also examines the challenges in institutionalizing GRB in local government, in relation to leadership and staff acceptance. It analyses the role of capacity building in raising awareness to have GRB accepted, i.e., the third sphere of cultural and behavioural change. Another challenge is setting up new structures (systemic change) to support GRB, at the local council and community levels; these aspects are relevant to the third and fourth spheres of power configurations.

8.3 Laying the Foundation

Penang did not underestimate the challenges in introducing GRB to the state nor did it attempt to rush the GRB proposal through. Gender advocates, especially from women's groups, and academics put in considerable effort into preliminary groundwork, given the openness of the then newly formed state government under Pakatan Rakyat,[5] which was in the opposition prior to the 2008 general elections. A starting point was the conference on 'Gender Mainstreaming: Justice for All' in 2010. It was organized by the newly formed non-government Gender Equality and

[5]Pakatan Rakyat is the People's Alliance, the opposition coalition consisting of three political parties.

Good Governance Society (3Gs),[6] Penang, and the Women's Development Research Centre (KANITA) based in the Universiti Sains Malaysia. This was the first time a specific call for GRB was made. One of the Conference recommendations was for the government of the State of Penang to adopt and implement GRB, linking its institutionalization in the state to the overall promotion of gender equality and good governance. Thus, the involvement of women's organizations and academia was a critical catalyst for the GRB project in Penang (Good Governance and Gender Equality Society 2011).

Positive feedback from the state government to this initial suggestion led to three GRB focused workshops, co-organized by 3Gs and KANITA, in 2011. These were endorsed by two key Penang State Executive Council (EXCO) members, one in charge of Local Government and the other in charge of Women, Family and Community Development (and Youth and Sports). Participants included key personnel from state and local governments. The first workshop, held in February 2011, was inaugurated by the Deputy Chief Minister of Penang (representing the Chief Minister who was away then). It was attended by 42 participants and introduced concepts, tools, methodologies and international experiences of GRB. An important outcome of the workshop was the formation of a GRB Task Force,[7] the first step in an explicit 'formalizing' of the Penang GRB initiative into structures involving State and local government partners. During this workshop too, a municipal councillor proposed the setting up of an ad hoc Gender and People with Disabilities Committee in his local council.[8] The second 'Advanced GRB Workshop' was conducted by world-renowned GRB expert Professor Rhonda Sharp in June 2011 and was attended by about 30 participants. The third workshop was held in November 2011, focusing on officers from an LA in Penang.

These workshops were considered crucial in building understanding of GRB and convincing state and local government representatives of the benefits of introducing it as part of local government practice. The workshops helped build capacity in gender issues, and generated greater awareness and interest in GRB. They also facilitated the emergence of key officials (including local government policy makers) who would help champion GRB in partnership with the local government, 3Gs and KANITA.

---

[6]While non-governmental in nature, 3Gs, registered in late 2009, was financially supported by the State EXCO of Women, Family and Community Development to facilitate education and awareness on gender equality. The idea then was for 3Gs to morph into a state women's body once it received sufficient traction and visibility. Thus, the ground work was already laid for cooperation between women leaders from the state government, academia and women activists in Penang.

[7]The GRB Task Force at this time comprised a representative (the Finance Director) of the MPSP, councillors from MPSP and MPPP and representatives from KANITA and 3Gs. Visits were made to MPPP and MPSP to learn about their budget cycle and to meet the leaders of the two LAs to brief them about the potential of GRB.

[8]After a year of lobbying, this committee was separated into two, with the Gender Committee (a new structure) being set up in January 2013.

The workshops generated optimism that GRB could merit serious attention if proposed to state and local governments. The GRB Task Force started developing a proposal (finalized and presented by a consultant in November 2011 as the project document or PRODOC) to present to the government. At the same time, a Scoping Exercise was conducted from July to October 2011 to look into the feasibility and challenges of introducing and implementing GRB into local government. As noted by the report:

> The Scoping Exercise took into central consideration the fact that the approach to mainstream GRB is about two major things. It is about making sure that the assessment of needs and service delivery is made with equal concern about women and men. This means, among other things, that the design, collection and use of data is appropriately institutionalised into the budget cycle to ensure the different needs and uses of women and men are equally acknowledged and valued. Secondly, it is an approach which insists that the processes of government, including budget planning, are opened up to include inputs from women and men at all levels of our society. This places GRB squarely in the context of the search for good governance (Lochhead 2011: Executive Summary).

As mentioned earlier, simultaneous lobbying was undertaken to encourage the Penang state government to establish a state-funded body to take charge of women's affairs. The Chief Minister finally approved this in mid-2011. In November 2011, the PWDC was established, although it began operations only in January 2012 (Ng 2012). Upon approval from the PWDC Board of Directors and members of the MMK[9] Women, Family and Community Development, a proposal was sent to the Penang state Executive Council to approve PWDC working with the two LAs in this project. The GRB pilot project, in partnership with the two LAs, became its first flagship project. All three organizations provided funds. The state and local governments would, as far as possible, be full partners from the beginning, an essential prerequisite for successful acceptance and institutionalization of GRB.

The challenge in translating a GRB pilot project into a fully institutionalized gender-mainstreaming effort was not downplayed. A number of potential obstacles were highlighted. However, optimism was based on a conjunction of factors and events which have rendered the State and Local Governments of Penang receptive to both the prospects of enhanced service delivery and inclusiveness that GRB offers.

Part of this optimism was based on the change in the State government that had occurred in Penang in March 2008: a momentous change that reversed trends of the past 50 years and found Penang with an opposition-controlled government. Its statement of intent included its ambitions to attain International City status and commitments to such concepts as people-oriented government and democratic participation, respect for diversity, equal opportunity and prohibition of discrimination.

The time was then ripe for the GRB project to commence. The care and time taken in lobbying and laying the groundwork for the pilot GRB project in Penang considerably increased the likelihood of local government acceptance and institutionalization of GRB. Two stages were important: acceptance first, followed over a

---

[9]MMK or *Majlis Mesyuarat Kerajaan*, the State Executive Council for Women, Family and Community Development.

period of time by institutionalization. Through the workshops, the GRB Task Force and the Scoping Exercise, key officers within the state and local governments had become aware of GRB. By late 2011, they were willing to commit their own funds to a pilot project. This was a huge step forward towards the ownership and initial institutionalization of GRB in the two LAs. At the same time, the lobby for a Penang state-funded body to take care of women's issues had borne fruit with the establishment of the PWDC with funding from the Penang state government.

8.4 Implementing GRB in Local Government: Challenges of Institutionalization

In a comprehensive matrix for a 3-year pilot project, the PRODOC set out five outputs that would govern the pilot project. These were:

1. An enabling and supportive environment within Penang Local Government (MPPP and MPSP or LAs) for GRB implementation and institutionalization.
2. GRB implementation of community pilot projects.
3. Development and establishment of a gender-disaggregated databank.[10]
4. Capacity development in GRB methodology and tools.
5. Increased public awareness and participation in budgeting processes.

The PRODOC clarified that each of the five outputs was intrinsically linked to the other. It set out how work must occur within the formal structures and processes of local government and informal and community settings. To create an enabling and supportive environment for the GRB project, new structures had to be created and new awareness and attitudes about gender had to be understood and internalized. GRB tools had to be learnt and applied to current budget planning and processes. The new participatory methodology introduced in the community pilots had to be internalized and accepted by the local council and the community. Thus, a process of organizational buy-in was needed with capacity building of council staff working alongside the PWDC project staff.[11] What were the issues and challenges? Table 8.1 depicts the lessons learnt since 2012 with respect to the SWOT (strengths, weaknesses, opportunities and threats) analysis put together by the Project Director.

[10]This was later amended to 'sex-disaggregated data used as an integral tool to better policy analysis and budgeting allocation'. A mapping of available sex-disaggregated data was conducted in July–August 2011. The consultant revealed that not all departments included the sex variable in their data collection. Another concern was the lack of specific and concrete data on the number of women and men using public facilities (e.g. gyms, swimming pools and libraries).

[11]The first year saw a three-person PWDC GRB project team of two project officers, headed by a Project Director and supported by two consultants and a GRB Advisor. In 2013, the PWDC team comprised a Project Director, four project staff, a consultant and the GRB Advisor. Contact persons were appointed in the two LAs to support the GRB pilot. There was no dedicated LA GRB staff.

Table 8.1 The 2013 SWOT analysis

Lessons 2013	Strengths	Weaknesses	Opportunities	Threats
MPPP/MPSP	Financial commitment	Theoretical commitment	Steering committee support	Time constraints
	High top management buy-in	Varying levels of participation	KRA/KPI driven	Competing priorities
			OBB	
Community	Openness	Perceived	Capacity Building	Varying levels of participation
	Eager to learn	'Gender' dilution	Widen activities	
	Willingness to take ownership	Own priorities	Multiple partnership	
			Community contracts	
Team	Great team spirit/teamwork	Substantive understanding	Continuous capacity building	High expectations
	High level commitment	Young in experience	International exposure	Workload

Source Penang Women's Development Corporation (2013)

Some of these lessons will be discussed in the following sections, particularly those related to institutionalization.

8.4.1 Organizational Buy-in

Initially, the two LAs had varying levels of acceptance and buy-in. As Kelleher/Rao (1999) noted, while an outside constituency is important (in this case, PWDC and the GRB team), the inside bureaucratic voice and strength is just as, if not more, important to ensure that new social practices are accepted. The top management was however very supportive. In one LA, GRB made an almost seamless entry into its newly adopted Transformation (*Transformasi* in the Malay language) programme that aimed to radically change the organization's behaviour. The *transformasi* journey indicated the Council's willingness to listen and respond to the people. It included an important change from a previous line-budgeting process to an output-based budgeting thrust, defined and driven by six Key Result Areas (KRAs), with accompanying strategies, action plans, activities and Key Performance Indicators (KPIs). Similarly, the other LA also worked on its KRAs/KPIs based on a key strategy, namely a Safer, Cleaner, Greener and Healthier Penang.

Although the leadership in the other LA was initially more sceptical, it became more accepting of the project. This was partly due to the commitment shown by several active councillors who supported the GRB team in reaching out to a community deemed difficult by council staff as the council owned and managed apartments occupied by low-income and underprivileged communities who had difficulties paying their monthly rental.

Understandably, the first year was rather slow as relevant implementers were just getting used to this new idea. GRB programmes and activities were competing with the existing priorities of council staff. In addition, GRB's participatory methodology was also new to the council staff who would have to work in and with the community. Gradually, changes occurred, with visible openness and eagerness from both communities to take ownership of their own stated priorities under this innovative process. This, in turn, influenced the LA officers to become more accepting of the GRB pilot despite their busy schedule. More financial resources were pumped into a low-cost flat when the LA saw first-hand that community mind-sets were changing.

Increased awareness of gender-based needs further prompted the LAs to take positive action on meeting the needs of different people. These included doubling/tripling of public toilets for women, setting up baby changing and lactating rooms, specially designed and designated areas for people with disabilities at public parks, and express lanes and special counters for the elderly and those with disabilities.

8.4.2 Creation of New Structures and Processes

New structures had to be created to direct and manage the project and organizational processes had to be invented. How could this new kid on the block be accepted as part and parcel of the organizational structures of local government? How could PWDC and LA staff be imbued with new knowledge, skills and resources to ensure the successful implementation of GRB? To achieve this, three new aspects were introduced. First, new GRB-relevant structures were created. Second, several capacity-building workshops were conducted, and third, two pilots were implemented at the community level.

8.4.2.1 GRB Structures

The new structures included a GRB steering committee at the highest level, GRB working committees in the two LAs and community structures in the two pilot communities (Fig. 8.3).

The GRB steering committee was chaired by the influential EXCO member in charge of Local Government and Traffic Management who had been very supportive from the beginning. Other members included the Presidents of the two LAs, the CEO of PWDC, the GRB Project Director, the GRB Advisor-cum-PWDC

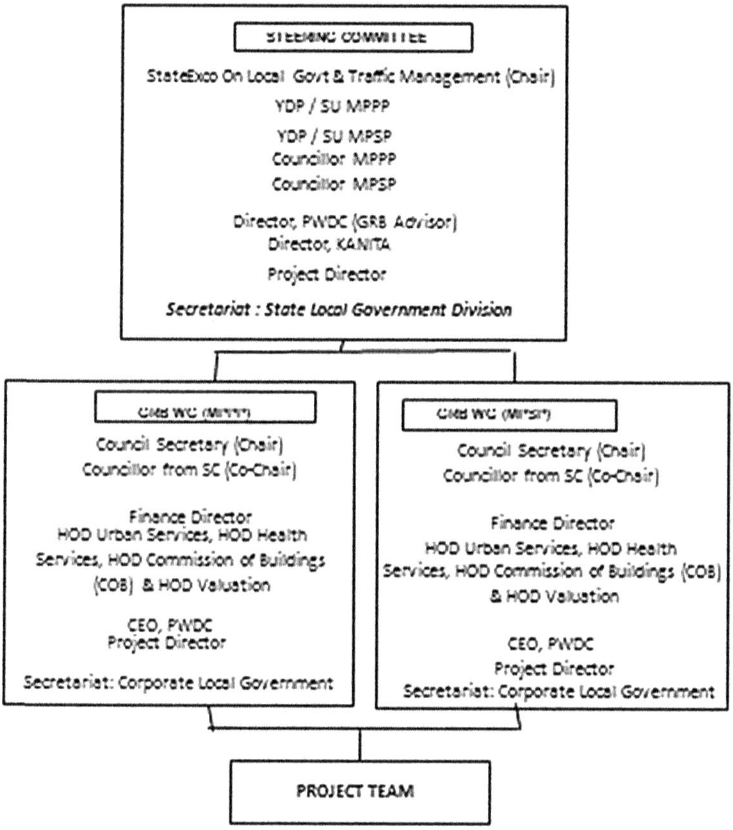

Fig. 8.3 GRB steering committee and working groups. *Source* Penang Women's Development Corporation (2013)

Director, and a councillor each from the two LAs.[12] The steering committee met twice a year and its role was to monitor the progress of the project and provide strategic guidance and direction to the project team. This structure worked well and proved to be important in legitimizing GRB in both state and local governments.

However, LA GRB working groups, often comprising heads of department, found it challenging to juggle priorities. As an alternative, technical committees supported by councillors were activated and became more viable. These committees comprised officers of departments directly responsible for acting upon decisions

[12] As in Fig. 8.3, YDP (Yang di Pertua) is the President of the local council. SU (SetiaUsaha) is the local council Secretary, while HOD is the Head of Department and COB is the Commissioner of Buildings.

within their jurisdiction. The other LA invited the GRB team to higher-level decision-making meetings.[13]

The third community structure was recently organized to facilitate a community contract between one of the local councils with the Residents' Association of a low-income community. It took almost six months to prepare both the institution and the community. The GRB team and the LA had to work creatively within the parameters of the standard operating procedures of the local government to introduce innovative approaches. For all those involved, it was a learning experience. This new modality meant that the residents' association could now, for the first time in council history, be set up as a sub-contractor to the local government and become custodian-cum-employer of the residents for the cleanliness programme (Fig. 8.4).[14]

8.4.3 Capacity Building: Changing Institutional Culture

Various capacity training workshops were organized for all stakeholders to help accommodate this change in work culture and to obtain new gender knowledge and analysis. Council staff seemed to be open to the idea of gender equality and women's roles in decision making. Research conducted in 2010–2011 found that the majority of the leadership in the two LAs concurred with the idea of gender equality and were not averse to women's leadership (Ng 2012a). The openness to change with respect to gender mainstreaming and gender equality would facilitate the introduction of GRB. However, it was still important that officers and councillors increase and deepen their knowledge and skills with respect to GRB.

A workshop, 'Piloting GRB in Penang', was held in April 2012 and facilitated by an expert from the Philippines. Its objective was to develop the technical capacity of stakeholders and PWDC-GRB project staff in terms of familiarity and competence in the key areas of GRB and planning processes, gender mainstreaming, gender needs analysis for GRB, community-level planning, budgeting and organizing, process documentation and budget dialogue. The expected learning outcomes were:[15]

1. Key LA officers, PWDC Project Team and other stakeholders/participants would acquire relevant GRB skills.
2. A pool of local GRB trainers would be created. They would be competent in achieving GRB awareness raising and in leading gender-aware appraisal of LA-level policies and services and other core areas.

---

[13]As of February 2014, the Gender Committee has taken over the responsibilities of the GRB Working Committee in one of the LAs to streamline decisions.

[14]As at end 2014 PPR Ampangan had also successfully established a similar community cleaning contract in their low-cost flats.

[15]In hindsight, the organizers realized that not all the outcomes were, or indeed could be, achieved in this rather ambitious workshop.

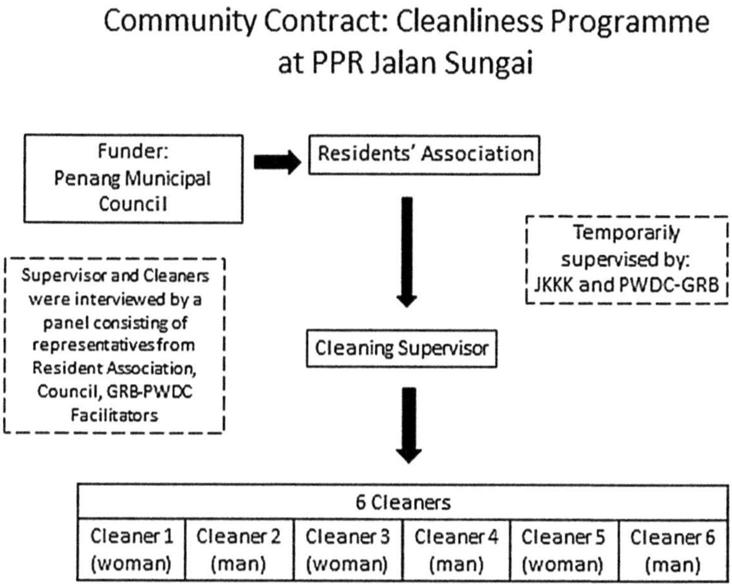

Fig. 8.4 Community contract structure. *Source* Penang Women's Development Corporation (2013)

3. A network of GRB experts in the state of Penang would be developed and strengthened.
4. LA Council management and staff would support integrating the gender perspective into local governance processes and would acquire relevant GRB skills.
5. Targeted community groups would acquire an understanding of GRB and relevant skills to provide inputs to LA Budgets.

This was followed by nine workshops, in 2012 and 2013, for various levels of officers and decision makers of the two local councils. The objective was to cover basic understanding of gender, linking gender needs to good governance. These workshops were generally well received. However, a workshop on GRB methodology and tools, planned in the PRODOC, was not conducted because international and locally trained expertise in this area was unavailable. The result was the inability to acquire skills to integrate gender into the budget cycle, including administrative procedures. This slowed down the institutionalization of GRB in local council planning, implementation and evaluation. As seen in Fig. 8.5, this output of 'heightening knowledge and skills in GRB methodology and tools' scored 50 %. One of the major challenges was getting targeted decision makers to attend training. Often, junior officers who attended training workshops felt planning and budgeting were beyond their decision-making spheres.

Despite the number of trainings conducted, uncertainties and differing interpretations of gender and GRB were evident even at the end of two years. Subsequently, the GRB team suggested to the Steering Committee, in October

OUTPUT 4	OUTCOME	ACHIEVEMENT	% SCORE	FLAG
Heighten knowledge & Skills in GRB methodology & tools	• A pool of local GRB practitioners acquire competence in GRB skills • Councillors, Council Management and Staff, integrate gender perspectives into local governance processes and acquire relevant GRB skills • Targeted communities and NGOs acquire understanding of GRB and relevant skills to effectively participate in LA budgeting • A Toolkit produced to help guide GRB implementation in the LAs	• 17 new Councillors & 4 HoDs are trained & more gender aware • New Councillors voluntarily join PPR events & activities • 100 level 27-48 trained at MPPP • 89 level 17-29 trained at MPSP • PPR Ampangan resident representatives attended MPSP's Budget Dialogue	50	

Fig. 8.5 GRB scorecard 2012–2013: Output 4. *Source* Penang Women's Development Corporation (2013)

2013, that after two years of implementation, institutionalizing GRB at the LAs needed to be scaled up. A major factor was inadequate capacity building of LA senior staff often due to competing priorities and appointment of new councillors at the beginning of the year.

This situation was confirmed by an international consultant who assessed the project in November 2013. In her report, she noted diverse views about 'gender'. The report observed that a typical statement was 'gender is (very) new to me'. Indeed some were relieved when they heard that gender would not be only about women 'against' men but include other differences like age (Frey 2013: 5).

It was thus heartening that the Steering Committee decided on a compulsory workshop for policy and decision makers comprising Heads of Departments and Councillors on GRB to ensure all groups were at par. This workshop, held in January 2014, emphasized the importance of gender sensitivity in work. A session introducing Outcome-Based Budgeting (OBB) with the necessary gender lens was also conducted. Seventy per cent of those who responded on the evaluation form rated the training 'average to good'. Many felt that they were able to better understand and articulate the meaning of gender and highlighted the need for more hands-on tools to apply gender perspectives to their work.

Planned training sessions for the rest of 2014 will cover more technical areas such as skills to interpret and integrate gender into the budget cycle, including GRB tools and analytical techniques especially for this group of policy and decision makers. These capacity-building sessions will, hopefully, change institutional culture, mainstreaming gender into the structures and processes of the two LAs.

8.4.4 Community Buy-in: Programmatic Interventions

An important 'output' was the implementation of community pilot services. Two themes driven by two external consultants were selected: cleanliness and safety. The pilot on cleanliness aimed at collaborating with the Local Agenda (LA21) by MPSP in a selected community on 3R (Recycle, Reuse and Reduce) activity. The safety pilot was to be conducted in two low-cost flats. Significant effort was expended in the first year on these two pilots. This section focuses on the learning experiences in the two low-cost housing projects.

The external consultant introduced a unique four-phase participatory methodology that resulted in the community buying into the GRB project, despite initial resistance and conflict.[16] A demographic survey was conducted as part of the first phase. Thereafter, residents were divided into five focus groups of different ages representing adult women and men (separately), those with disabilities, and young girls and boys (separately as well) to discuss their needs and concerns, which were then prioritized. A voting exercise, using paper money, was then conducted over three days. The result was that one community voted for cleanliness while the other voted for recreation, the implementation of which formed the fourth and final phase of this methodology. Other programmes for the community and for women were also conducted in between these phases (which lasted for nearly two years) to create solidarity. To ascertain the impact of such activities, GRB tools such as Benefit Incidence Analysis and Beneficiary Impact Assessment were then undertaken together with the community, including women, to assess their responses to the pilot.

This participatory methodology and the ensuing visible outcome led to the GRB pilot being accepted by the two communities. Concrete results emerged as something of a surprise to the project team which was unsure of how, for example, the recreation park—designed together with the residents and local council—would be negotiated and built. In the other low-cost apartment complexes, despite bureaucratic red tape, a cleaning contract was negotiated with the residents who have now taken over the cleaning contract (and security later on) for their community. All this led to further buy-in of GRB by the two LAs.

8.5 Conclusion

It is evident that the PWDC and the two LAs have achieved a lot in two years. The Project Director presented the following score card to the Steering Committee, articulating the success rate of each output (Fig. 8.6). The score card on the first output, institutionalization, showed 50 % achievement. Figure 8.7 provides further details of specific outcomes and achievements.

---

[16] See Shariza Kamarudin, this volume, for details of the pilot in the two communities.

OUTPUTS		%
1	Creating a supportive and enabling environment for implementing & institutionalising GRB at the Local Authorities (LAs) in Penang	50
2	GRB implementation of community pilots	80
3	Use of sex-disaggregated data to a better policy planning & budgeting	50
4	Capacity building in GRB tools & methodology at the Las & community	50
5	Increased public awareness & participation in budgeting processess	80

Fig. 8.6 GRB scorecard 2012–2013: Outputs 1–5. *Source* Penang Women's Development Corporation (2013)

OUTPUT 1	OUTCOME	ACHIEVEMENT	% SCORE	FLAG
Supportive & Enabling Environment to implement GRB	• Increased understanding and acceptance of GRB at all levels of Local Government, among the Councillors and within the community • Supportive structures established within the LAs & the implementation of GRB over the long-term sustained	• Management Steering committee formed • GRBWG formed • Gender Committee at MPSP formed • Top Management commitment • Top Management participation • GRB exhibit at IWD/Hari Majlis • Budget Dialogue conducted by MPSP • Budget on-line survey conducted	50	

Fig. 8.7 GRB scorecard 2013: Output 1. *Source* Penang Women's Development Corporation (2013)

The rating on institutionalization was not unexpected. Other countries have taken several years to firmly establish GRB in their planning and implementation processes. For Penang, the first two years saw major efforts put into Outputs 2 and 5 to achieve a participatory approach that would translate into considerable impact—tangible and visible to the LAs and the public at large. Traditionally, change is initiated from inside the organization. However, the project team added another synergy—change from the outside, which was then articulated in a dynamic and often uneven, if not risky, manner with structures and processes within the institution. The end result was the successful implementation of the recreation area and the community contract. The latter was a breakthrough of sorts that spoke volumes for this participatory approach. A lot of interest was garnered and GRB easily became a buzzword despite the lack of a real understanding of the term (Frey 2013).

Given this wave of change and empowerment from the outside, institutions are challenged to move in tandem. To strengthen this move, work must now come from within formalized institutional and organizational spheres (Kelleher/Rao 1999). Structural changes must be instituted through policy and procedures. Greater emphasis must be placed on instilling ownership, and stronger political will is needed to achieve the next wave of change that will fashion a truly supportive and enabling environment.

8.6 The Way Forward

Changing practices, rules and procedures and internalizing new values, visible or invisible, towards a gender responsive and participatory type of budgeting is a long-term endeavour. For Penang, such changes were facilitated by external advocacy in the earlier stages, supported by an empathetic leadership (that provided the resources) and an open institutional culture. The innovative participatory methodology, albeit a lengthy process, showed the two LAs the possibility of community engagement and empowerment. This was facilitated by a new state government that considered good governance and civil society participation integral to its reformist credo. The Penang pilot argues that a good, sustainable and inclusive GRB programme must have both critical aspects—community involvement and GRB institutionalization in government. The crucial gap to fill now is further institutionalization of GRB to achieve the overall goal of good governance and gender equality (Fig. 8.8).

As the international consultant noted:

> In the short span of 2 years the GRB team has done very well. GRB is well known to many stakeholders with the local administration and there are formal and informal networks and bodies to implement GRB. Since GRB is a process which requires transforming procedures and cultures within organisation, it is not something that can be implemented in 2 years (Frey 2013: 17).

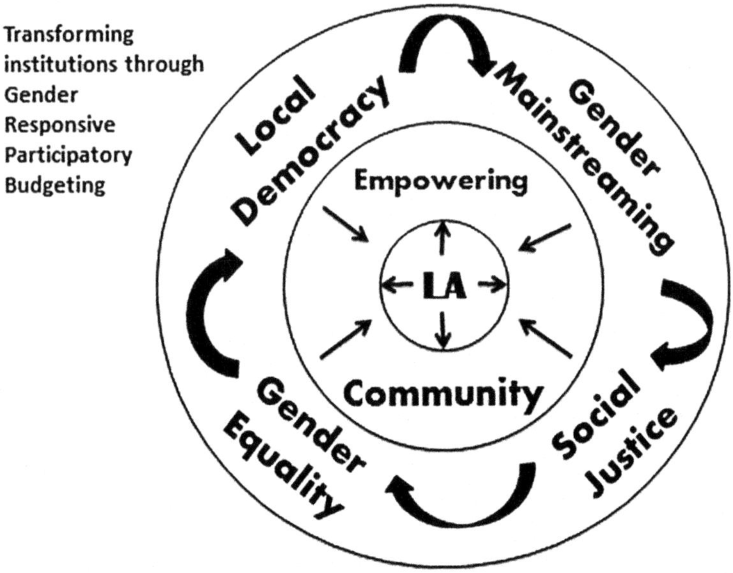

Fig. 8.8 Transforming institutions through GRB. *Source* Penang Women's Development Corporation (2013)

Much remains to be done in terms of institutionalization of gender in local governing agencies, but the process has been set in motion and with continued commitment from both LAs and the state it will certainly succeed. What is the way forward then? What would the next cycle look like and how does one move ahead? Several recommendations are provided to deepen the institutionalization process.

Firstly, it is important to have a policy framework that legitimizes GRB as a state and local government strategy.[17] PWDC is finalizing a gender policy for the state that would gender mainstream the state, including GRB processes. With GRB legitimized through state (and LA) policy, integration of gender into the budget cycle can be enhanced and deepened. A policy framework will facilitate a holistic approach and commitment towards gender mainstreaming of policies and practices of good governance. Without a clear policy, GRB will remain an additional activity on the fringe of discussions. It can be easily sidelined due to competing priorities. Thus, the danger of GRB remaining a pilot project continues. The funds currently provided to PWDC are a separate, if not isolated, women's fund, making it difficult for gender to be integrated holistically into the planning and implementation of policies and budgets. In this context, budget speeches at year-end need to include a GRB statement to enable a smoother integration in the next project cycle.

[17]In 2009, under Article 13(3) of the Constitution, Austria made gender budgeting legally binding (Klatzer et al. 2010).

Secondly, funds and human resources must be committed towards GRB and the two LAs must take greater ownership of GRB in their respective programmes. At present, PWDC facilitates GRB-related projects for the LAs. The more arduous journey of transforming administrative and standard operating procedures and work cultures now becomes critical. This would include gathering systematic and holistic sex-disaggregated data to support these transformational processes. Thus, an extended organizational buy-in is imperative in the next stage. It would help if a dedicated gender unit could be set up and entrusted with gender-related, including GRB, planning and implementation in partnership with the PWDC project team. Overall, this means that the two LAs should continue their financial commitment to the next three-year cycle and provide the necessary human resources to move GRB forward.

Thirdly, such institutionalization must be accompanied by increased capacity building for staff. A comprehensive strategy on capacity development must be developed so that regular trainings and workshops can be organized to enhance the technical skills and knowledge of LA staff on gender and GRB. The GRB Steering Committee has agreed that capacity development on GRB will form part of the compulsory annual seven-day training for LA staff. With this knowledge, GRB indicators can be included in the KPI of LAs.

Only through such steps will gender and participatory budgeting be a recognized and legitimate strategy, contributing towards gender inclusivity and sustainability of a Penang people-centred government. Only long-term political will and sustainable commitment will ensure the most supportive and enabling environment to transform the mainstream to achieve gender equality and social justice in Penang.

References

Frey, Regina, 2013: *Evaluation Report. PWDC's Gender Responsive Budgeting Project in Penang* (Berlin: Genderbüro).

Helmke, Gretchen; Levitsky, Steven, 2004: "Informal Institutions and Comparative Politics: A Research Agenda", in: *Perspectives on Politics*, December, Issue 4: 725–740.

Illo, Jeanne Frances I.; Encinas-Franco, Jean; Villaseñor, Jon Michael R.; Leyesa, Maria Daryl L., de los Trino, Frances Chariza I., 2010: *Accounting for Gender Results. A Review of the Philippine GAD Budget Policy* (Manila: Women and Gender Institute, Miriam College).

Kelleher, David; Rao, Aruna, 1999: "What is Gender at Work's Approach to Gender Equality and Institutional Change?"; at: http://www.genderatwork.org/Portals/0/Uploads/Documents/Gender-Equality-and-Institutional-Change.pdf (15 October 2015).

Klatzer, Elisabeth; Schratzenstaller, Margit; Buchinger, Birgit; Schaffer, Nicole, 2010: "Gender Budgeting in the Constitution—A Look at Formal and Real Conditions in Austria", in: *Internationale Politik und Gesellschaft*, 2: 48–64.

Levy, Caren, 1996: *The Process of Institutionalising Gender in Policy and Planning: The 'Web' of Institutionalisation*. Working Paper No. 47 (London: University College London).

Lochhead, James, 2011: *Scoping Exercise Report*. Good Governance and Gender Equality Society (Penang: PWDC).

Ng, Cecilia, 2012a: "Gender and Governance: The Politics of Federalism in Malaysia", in: *Kajian Malaysia*, 30,2: 1–26.

Ng, Cecilia, 2012b: "Right Time, Right Place and Some Luck: PWDC in the Making", in: *Penang Monthly*, April: 18–20.

Penang Women's Development Corporation, 2012: "Project Document (PRODOC): Gender Responsive Budgeting for Gender Equality and Good Governance in the Local Government of Penang: Proposal for a Three Year Pilot Project" (Penang: PWDC).

Penang Women's Development Corporation, 2013: "Report to the GRB Steering Committee", October (Penang: PWDC).

Chapter 9
Institutionalizing Gender Responsive Budgeting in National and Local Governments in Nepal

Purusottam Nepal

Abstract The chapter discusses how gender responsive budgeting has been introduced and implemented in Nepal. It looks at the changes at the national level in terms of structures and processes. The chapter also examines how changes have been made at the local level to incorporate gender sensitive participatory governance through the Local Governance and Community Development Programme. The conclusion delves into the challenges faced in this journey of combining gender responsive and participatory budgeting in Nepal.

Keywords Gender budgeting · Participatory governance · Local government · Accountability · Monitoring mechanisms · Grassroots women

9.1 Introduction

This chapter begins by describing how gender responsive budgeting (GRB) was formally introduced in Nepal in fiscal year (FY) 2007–2008 through the initiatives of the Ministry of Finance. It then discusses national-level changes in budgeting structures and processes that facilitated GRB implementation. These include institutionalizing a GRB committee to prioritize the needs of women; introducing a gender code classification system for programmes and projects; using sex-disaggregated indicators and an outcome-monitoring mechanism. A policy of minimum 33 % women's representation in various grassroots institutions and 10 % allocation to women's programmes from the local budget have also helped to accelerate the GRB process.

The chapter also examines the experience of the Local Governance and Community Development Programme (LGCDP) in attempting to promote gender sensitive participatory governance. It notes that in the first phase (2008–2012), LGCDP created some 40,000 grassroots institutions with about 50 % women's

---

Purusottam Nepal, Under Secretary/National Programme Manager, Local Governance and Community Development Programme (LGCDP), Programme Support Unit, Ministry of Federal Affairs and Local Development, Kathmandu, Nepal. Email: nepalpuru@yahoo.com.

C. Ng (ed.), *Gender Responsive and Participatory Budgeting*,
SpringerBriefs in Environment, Security, Development and Peace 22,
DOI 10.1007/978-3-319-24496-9_9

membership. It covers the achievements and challenges in capacity building (including technical expertise on gender responsive planning, budgeting, monitoring and evaluation) and changing attitudes. In conclusion, it acknowledges the important role of GRB as a tool at the local level to ensure citizens' participation in policy formulation and decisions, as well as the accountability of local governments in delivery of equity and the economic and social rights of women and men.

9.2 Gender Responsive Budgeting in Nepal

GRB is a tool to integrate the gender perspective into every step of the budget process, i.e., planning, implementing, monitoring and evaluating. It is directed towards the needs and interests of women and men from different groups and ensures that these needs are addressed in public budgeting. It also ensures that budget policies take into consideration gender issues in society such as gender-based discrimination. GRB is a tool to mainstream gender in policies and plans, to redress inequalities and to promote women's economic, social and political rights. It is also an extension of the concept of performance budgeting, which focuses on the achievement of results in the form of outcomes as distinct from performance measured merely in terms of budgetary expenditure.

GRB initiatives seek to create an enabling policy framework, build capacity and strengthen monitoring mechanisms to support accountability. Their main goal is to create awareness among the public regarding gender issues and make governments more responsible for drafting, implementing and updating policies and budgets related to gender issues. Sharp (2003) has categorized GRB outcomes as follows: (1) to raise awareness and the understanding of gender issues and increase the impact of budgets and policies; (2) to make governments accountable for their budgetary and policy commitments to gender equality and (3) to change and refine government budgets and policies to promote gender equality. Correspondingly, gender responsive budget can be seen in two ways: (i) To assess the impact of government expenditures and programmes on the economic and social position of women and men; (ii) To develop strategies that will result in a more gender sensitive allocation of resources.

GRB also ensures availability of and access to public services with the objective of moving towards a more inclusive development. Usually, women and marginalized groups have diverse needs that are often neglected or not addressed. GRB identifies such needs, leading to interventions that address the gender and socio-economic gaps in policies, plans and budgets. Hence, it is critical for investigating and monitoring the extent to which public policy, programmes, budgets, aid modalities and expenditure patterns are gender responsive.

Within a framework of overall reform in budgetary processes, Nepal has adopted GRB to fulfil its constitutional requirement and international commitments to gender equality under the Convention on the Elimination of All Forms of Discrimination against Women (CEDAW), the Beijing Platform for Action (BPFA) and Millennium Development Goals (MDGs) since FY 2007–2008. Public budgetary expenditure has been classified as directly responsive to gender, indirectly responsive and neutral across all sectors. Efforts have been made to institutionalize gender mainstreaming through GRB across all sectors and levels of governance. The outcome of GRB initiatives can be seen in Nepalese society in general and specifically in State and non-State institutional mechanisms and in various developmental activities.

9.3 Institutionalizing GRB: Reforms in National Budgeting Structures and Processes

The 2007 Interim Constitution of Nepal, through its preamble, obligations, directive principles and policies of State, has accepted State restructuring, decentralization and devolution of power as a national policy framework to address gender mainstreaming and inclusion in national development. It provides that all socially, economically and geographically disadvantaged groups (DAGs), sectors and communities should have access to social, economic, cultural and political rights, opportunities and social security. Subsequent periodic plans have been in line with the Constitution of Nepal with an emphasis on gender sensitive, equitable and inclusive development.

Earlier on, the Tenth Plan (2002–2007) had already focused on women's empowerment, reinforcing gender equality, formulating a gender mainstreaming policy and adopting a GRB system. Consequently, Gender Focal Points in each ministry and the National Commission for Women were established. A study on 'Gender Budget Audit in Nepal' was conducted in 2003 and from 2003 to 2008, gender assessments and gender budget audits were conducted by seven ministries—Agriculture, Forest and Soil Conservation, Women, Children and Social Welfare, Education, Local Development, Health and Labour, and Transport Management.

The Interim Plan (2007–2010) adopted a policy of engendering development through economic, social and political processes to ensure inclusive development. The 3 Year Plan (2010–2013) provides for GRB institutionalization at the central level and for replicating it in local planning, programming, budgeting and monitoring.

In terms of GRB, the Government of Nepal initiated GRB institutionalization by forming a Gender Responsive Budget Committee (GRBC) in 2005 within the Ministry of Finance (MOF) to guide the process and give it continuity. The GRBC developed five criteria for evaluating the gender responsiveness of government programmes and projects: women's participation in programme formulation and implementation, benefits accruing to them, capacity building, contribution to women's employment/income generation and reduction in women's workload and qualitative improvement in their time use. In the same year, a Gender Management System (GMS) was established in 2005 within the National Planning Commission for institutional reform.

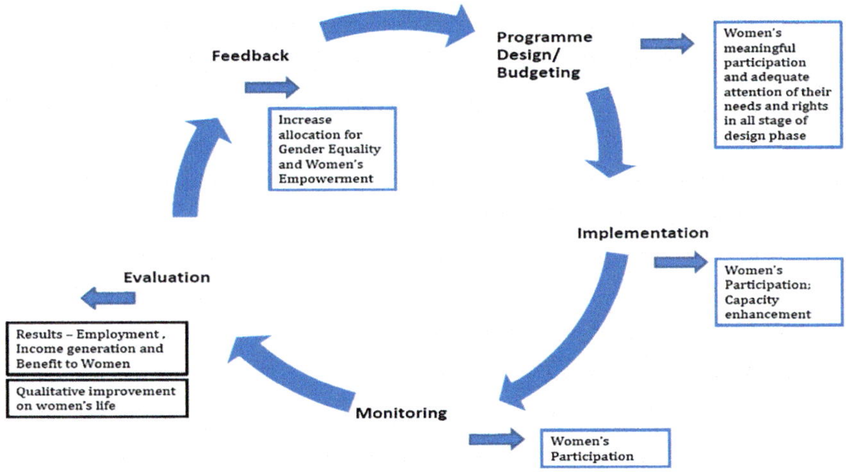

Fig. 9.1 GRB institutionalization process in Nepal. *Source* Ministry of Finance (2012)

As a result, the adoption of GRB has brought resource allocation for gender equality and women's rights to the forefront, strengthened gender mainstreaming mechanisms within government and increased budget allocations for women-specific needs such as combating violence, safe motherhood and scholarships for girls. In Nepal, GRB is institutionalized directly into the national budget cycle as shown above. Figure 9.1 shows the programming, implementation, monitoring, evaluation and feedback cycle for GRB in Nepal and the expected results from this intervention in all steps.

As a result, the central government's annual budgeting and development programme documents have been streamlined to reflect GRB. Accordingly, detailed information required for the review of budgetary allocation and a study of GRB outcomes are included in national budget documents, annual programme documents and periodic plans.

9.4 Criteria for GRB Classification

Computer software for GRB classification and categorization, such as Line Ministries' Budget Information System and Budget Management Information System were introduced from FY 2007–2008. They were based on the following three criteria:

- Directly responsive—if government-funded programmes are beneficial to more than 50 % women;
- Indirectly responsive—if government-funded programmes are beneficial to 25 to 50 % women; and

Table 9.1 Gender responsiveness indicators

SN	Indicators	Percentage
1	Women's capacity building	20
2	Women's participation in formulation and implementation of programmes	20
3	Women's share in benefits	30
4	Support to employment and income-generating activities for women	20
5	Quality reform in time use and minimization of women's workload	10
	Total	100

Source Ministry of Finance (2012)

- Neutral—if government-funded programmes are beneficial to less than 20 % women.

The classification was based on the indicators in Table 9.1. The above indicators provide guidelines for the allocation of the government budget in gender responsive programmes for different sectors. However, the basis for determining what is directly and indirectly supportive of gender-related and gender neutral programmes and activities is not clearly defined in the budget speech and Red Book prepared by the MOF and Annual Development Programmes prepared by the National Planning Commission. This is because the categorization of budget and programmes is ad hoc. The budget allocated to the Ministry of Women, Children and Social Welfare as well as the local government's budget in any gender responsive category is categorized as gender budget even if the money is spent on non-gender expenses. This lack of clarity has left a significant impact on women, as a budget that seems to be gender responsive often does not work for women's empowerment. To overcome this ambiguity, detailed GRB guidelines have been issued.

9.5 Review of Budget Allocations for Women

A broad review revealed that in various sectors, budget allocation in the directly supportive category is increasing annually, while it has been decreasing consistently in the neutral category since FY 2007–2008.

Table 9.2 also shows that the percentage of budgetary allocation for programmes directly benefiting women has increased gradually, from 11.3 % in FY 2007–2008 to 21.75 % in FY 2013–2014. Neutral budget allocation decreased from 55.54 to 34.31 % during this seven-year period. However, the budget allocated for women is relatively low compared to the actual requirement for gender equality. The amount of money allocated to women from excluded groups and the amount of expenditure thereon is also not indicated. As a result, money used for building roads was categorized as a gender budget, and money used for building community halls was

Table 9.2 GRB, 2007–2008 to 2012–2013. NRs in billion

SN	FY	Directly supportive		Indirectly supportive		Neutral	
		Amount	Percentage	Amount	Percentage	Amount	Percentage
1	2007–2008	19.01	11.30	56.03	33.16	93.87	55.54
2	2008–2009	32.91	13.94	83.58	35.41	119.53	50.64
3	2009–2010	49.46	17.30	104.16	36.43	132.32	46.27
4	2010–2011	60.61	17.94	112.65	36.30	154.64	45.76
5	2011–2012	73.33	19.05	176.21	45.78	135.35	35.17
6	2012–2013	87.07	21.51	178.63	44.13	139.11	34.36
7	2013–2014	112.5	21.75	227.3	43.94	177.4	34.31

Source Budget Speeches (various issues), MOF; District-wise Development Programme and Budget Allocation (various issues), NPC

categorized for disadvantaged groups. A computer centre was set up with funds specified for Dalit women who were illiterate, which ultimately only benefited the village elite (see Box 9.1 and 9.2).

Box 9.1: Case Study 1: Roads for Women

A major portion of the budget for women and children went to school construction and furniture (30 %), health post support (21 %), and road construction (14 %), as was the budget allocated for Dalits and Janajatis. Of the 13 drinking-water projects, ten were funded from the budget allocated for women and children and the remaining three from the budget for Dalits and Janajatis. Not a single drinking-water project was funded from the general budget of the Village Development Committee (VDC). In Paudeshwar, Aurah, and Dhalkebar VDCs, the entire budget was spent on road construction. In Etharba VDC, the whole amount was used for electrification. This shows that the target groups have very little say and influence over the way funds meant for their benefit are actually used.

Box 9.2: Case Study 2: Political Misdirection Derails Women's Benefits

In Bhatkhola VDC, Syangja district, the budget allocated for the women's programme was frozen for 2 fiscal years. Then in FY 2010/2011, the budget for women was allocated to a computer training centre by an influential member of a political party. Women from the mothers' and forest user groups, who were present at the project selection meeting, clapped with the rest though the computer training centre would not benefit them. The proposal passed. "What else could we do?" asked one of the women at the meeting, conceding that they lacked the political savvy and ability to offer resistance (LGCDP 2012).

This realization led to the following interventions: (i) Developing capacity of government machinery at the national and local levels; (ii) Establishing GRB focal points in relevant sectoral ministries; and (iii) Organizing a series of training programmes and workshops on GRB. With these actions, the situation is gradually improving.

9.6 GRB Interventions and Results

Efforts have been made to enhance the capacity of the government machinery to understand and deal with gender mainstreaming, particularly in development activities in agriculture, local governance, education and the police force. As discussed earlier, gender focal points have been established in all ministries and major departments. Notable institutional reforms in education and health sectors include the decentralization of management functions to District Development Committees (DDCs), Village Development Committees (VDCs) and community-level institutions. A series of gender sensitization training programmes and workshops have been conducted for all levels of government officials. Gender audits of many ministries have been conducted and recommendations are being implemented. Table 9.3 highlights GRB interventions and the results achieved, from 2007–2008 to 2011–2012. As is evident, women's involvement in decision-making processes and other public spheres is gradually increasing. However, GRB benefits such as greater equity through improved policy coherence, institutional reform and capacity development are yet to be realized.

9.7 GRB Institutionalization at the Local Level

The Local Self-Governance Act (LSGA), 1999, and its regulations have provided two layers of local authority: VDCs (3,915) and municipalities (58) at the local level in rural and urban areas respectively, and DDCs (75) at the district level in rural areas. Nepal is committed to a decentralized governance system and gender mainstreaming has been one of its major policy objectives in local governance. The LSGA has mandated women's participation in local governance institutions. As a result, 20 % women's representation is mandated at the ward level; though at higher levels only a nominal representation is mandated. The government has mandated a minimum 33 % women's representation at all levels through a bill in parliament.

In 2008, the Ministry of Local Development (now the Ministry of Federal Affairs and Local Development [MoFALD]) adopted the Local Body Gender Budget Audit Guideline. MoFALD also issued a Gender Equality and Social Inclusion (GESI) Policy and Strategy for the local level in 2009 with provisions for

Table 9.3 GRB interventions and achieved results

	Objective	Action	Results
1	To establish women's rights on land	Land registration fees reduced by 25 % in municipalities, 30 % in VDCs and 40 % in hilly areas if ownership certificate is in women's names	Women's land ownership increased to 23 % in 2011–2012 from 8 % in 2001–2002; equal rights for sons and daughters over ancestral property
2	At least one third of all public positions to be occupied by women	Quota system for new vacancies	There were 57 women in the 265-member Nepal Parliament, 197 women in the 601-member Constitutional Assembly (quota system secured 33 % for women), 13 % women in government service in 2011–2012 (up from 8 % in 2001–2002). The number of women in the army and police increased (women's cell, women's barracks and separate toilets established), 20 % seats were set aside for women in local bodies and 33 % in all committees; preparatory classes are held for women appearing for public service examinations
3	To decrease maternal mortality rate towards achieving MDGs	24-hour maternity services in all government hospitals, money for transportation expenses to get to government hospitals, and cash grants for four regular check-ups before delivery	In 2011–2012, maternal mortality rate fell to 229 from 510/100,000 in 2001
4	To stop violence against women	Safe houses for victims of domestic violence and sexual harassment, fast-track courts, one-stop crisis centres, safe houses for victims of trafficking	2010 was declared as the year to end gender-based violence (GBV). The government has prepared a five-year national strategy on violence against women and allocated funds of NR 10 million for victims of GBV and NR 10 million for rehabilitation of trafficking victims
5	To decrease dropout rate of adolescent girls	Compulsory girls' toilets in public schools and scholarships	School dropout rates among girls fell to 5 % in 2011–2012 from 16 % in 2001–2002 due to toilet facilities; women's literacy rate rose to 56 % from 42 % in 2001–2002 and enrolment rate in primary schools was at 96 %

(continued)

Table 9.3 (continued)

	Objective	Action	Results
6	To give women priority in school teacher, village health volunteer, development organizer/co-planner and social mobilizer roles	45 % recruitment of women on the basis of an inclusive system	Women accounted for 29 % of the teachers in government schools and women's participation in all community-level organizations was almost 50 %
7	To increase women's employment and participation in economic activities	10 % rebate on income tax to women professionals, 20 % tax exemption to companies employing 100 persons of whom 33 % are women, women entrepreneurs' village fund (NR 10 million) and self-employment programme for 45,893 women	Women employees increased in departmental stores, women entrepreneurs increased in micro and small-scale businesses and informal sector manufacturing, and 2,000 cooperatives were operated by women.
8	To increase women's participation in development activities	Extend women's development programme in all VDCs gradually	Women's empowerment and gender equality programmes were extended to 3,665 VDCs out of a total of 3,915
9	To develop Nepal as a unique country in South Asia with respect to GRB institutionalization	Establish Nepal as an example for GRB application in South Asia	All development partners should explain how their contributions directly or indirectly affect GRB on Aid Management Information System

Source Reports of various sector Ministries and Office of the Prime Minister and Cabinet Secretariat from FY 2011–2012

structural and programmatic improvement. Specific interventions at the local level were:

- Mandatory women's representation, ranging from a minimum of one to 33 % in all committees such as Integrated Plan Formulation Committees (IPFCs) and Supervision and Monitoring Committees (SMCs), Ward Citizen Forums (WCFs), and Users' Committees (UCs);
- Priority for programmes addressing gender and social inclusion when allocating development budgets;
- Mandatory provision of 10 % budget allocation by local bodies (LBs) for women's programmes, 10 % for children's programmes and 15 % for DAGs and other targeted communities;
- Equal wages for women and men;
- Developing specific implementation and monitoring directives incorporating gender and inclusion perspectives; and
- Establishing Women's Development Offices (WDOs) at the district level and appointment of Gender Focal Persons in all line agencies respectively.

Some of the interventions are discussed in the following sections.

9.7.1 Women's Participation in Local Planning/Programming, Budgeting and Project Implementation

Government policies, directives and guidelines and newly established grassroots institutions related to gender mainstreaming have had a positive impact on local governance. Efforts are being made to involve women in district, municipal and village level planning, although representation does not always reach 33 %. Women's workloads and other time constraints also mean that they cannot participate as frequently as men can. With no financial incentives, women from the poorest communities are unable to participate during working hours. Subsequently, local committees are now being encouraged to schedule appropriate meetings to ensure greater participation from women.[1] WCFs are grassroots institutions which act as social mobilizers.[2] They have helped to ensure 33 % women's representation in all committees at the local level, including decision-making positions. Consequently, local community leaders have encouraged women to participate in meetings, by changing the timings to their convenience. The incentive for participation is institutional recognition of their decisions in the village's planning process. Different organizations representing various groups, including women, are regularly invited to DDC, municipality and VDC councils where annual plans, programmes and budgets are approved.

9.7.2 Budget Allocations and Utilization at the Local Level

Budget analysis for the GRB initiative shows that the operating procedures of LBs have led to increased budget allocations for women-related programmes. Some DDCs have initiated gender budget studies and gender-focused programmes. Orientation training for GRB and Gender Audits has been initiated by many LBs for a broad group of actors, including local government officers, representatives of decision-making bodies, members of community-based organizations (CBOs) and non-governmental organizations (NGOs).

[1]The LGCDP has put one, usually female, social mobilizer in each VDC and each ward of municipalities. There are altogether 4,600 social mobilizers in LGCDP. These social mobilizers facilitate meetings, support poor women to raise their voice, help to record minutes and give trainings on the reflect model of social mobilization. The women members of the WCF/CAC said that they have been empowered through this model of social mobilization. Before joining the groups, they had hardly spoken during meetings, but now they are raising their voices and taking a stand until male members agree to reasonably address their issues. Indeed, it has been a successful model of empowering women on the process of local planning and accountability.

[2]After sitting in reflect classes for one year, all the members of CAC/WCF are now working as social mobilizers.

Similarly, analysis of VDC budgets indicates an increased budget allocation and expenditure in programmes for targeted groups including women. This appears to be due to the provision of mandatory women's participation in decision-making processes and allocation of specific funds to address the needs of women, children and other DAGs. The need for compliance as stipulated in the Minimum Conditions Performance Measures (2008) has also contributed to institutionalizing GRB in LBs. Indeed, GRB initiatives on the supply side have focused on improving knowledge and awareness of district government officers and improving their ability to understand and implement relevant laws, policies, guidelines and regulations such as LSGA, Gender Equality and Social Inclusion Policy (2009), Gender Responsive and Social Inclusive Budget Formulation and Audit Guidelines (2012) and Local Resource Mobilization and Management Guidelines (2012). Undoubtedly, all these measures have contributed to increased budgetary allocation by LBs to address the needs of women, children and DAGs.

9.7.3 Institutional Arrangements

At the central level, MoFALD established units in 2009 to mainstream GESI in its plans, programmes and activities. These units conduct different activities including GESI-sensitive amendments and reforms in policies and strategies, preparing manuals and frameworks, developing capacity, coordinating and consulting for effective GESI mainstreaming at various levels. At the local level, GESI implementation committees have been created in all DDCs with mandates to review and analyse district annual plans, budgets and expenditure from the GESI perspective and to develop GESI capacity of stakeholders.

All line agencies, federations, networks and NGOs working on GESI issues in the districts are to be represented in the committee. At the grassroots level, WCFs and Citizen Awareness Centres (CACs) have been formed in all VDCs and municipal wards as an entry point for all development activities including GESI. WCF members are nominated by the people through a mass gathering; various social groups (both 'elite' and DAGs) are represented. WCFs play a major role in planning and monitoring development activities. CAC members are mostly from among Dalits, Janajatis and DAGs, including women. Such social mobilization policy guidelines ensure that excluded groups have a voice in committees, while active local leaders and social mobilizers at VDCs and municipal wards have also helped this process. As of July 2013, 33,166 WCFs and 4,082 CACs had been formed.

Social accountability mechanisms in LBs not only institutionalize GRB but also develop citizens' capacity to ensure easy and equitable access to public goods and services. These include: (i) LB grant expenditure and community engagement survey, (ii) public hearing and compliance monitoring, and (iii) community-based monitoring of local government/public goods and services. Parameters related to GRB institutionalization have been included in the results-based monitoring system. This provides information on women's participation in local-level planning, number and cost of projects demanded by women, children and DAGs, and number and cost of projects included in LBs' annual plan in favour of women, children and DAGs. The system also provides information on women's participation in LB councils, public audit and public hearing activities. Moreover, it also provides information on the number of beneficiaries in terms of sex and ethnicity. The GESI policy highlights six dimensions of exclusion and inclusion: gender, geography, caste, ethnicity, age and ability. The monitoring and evaluation system, established two years ago, provides disaggregated information on these dimensions.

9.8 LGCDP (Phase I)—Evaluation of Grants and GRB Projects

The Local Governance and Community Development Programme (LGCDP) is a joint initiative of the Nepal government and donors for effective local governance (Ministry of Local Development 2008). It provides an overall framework for strengthening decentralization and devolution and establishing an improved local governance system for effective delivery of basic services and the empowerment of citizens, especially women, children, DAGs and their institutions. The strategic goal of LGCDP is to promote inclusive local democracy through local community-led development that enables the active engagement of citizens from all sections of society. LGCDP supports socially inclusive and gender-friendly approaches in the design and implementation of programmes at the local level. Affirmative policies have been introduced in favour of the poor, women and DAGs to maximize their participation in and benefits from the programme. Procedures have been designed to ensure greater equity and efficiency while working with communities and for targeting the poorest and most disadvantaged segments of these communities. To assess the effectiveness of this first phase of LGCDP, various studies have been conducted by MoFALD and UN agencies as well as the local district committees.

MoFALD has a monitoring and evaluation (M&E) framework for different activities including GESI. The results-based monitoring system has captured GESI-related activities. Moreover, MoFALD emphasizes perception surveys to assess the impact of GESI-related activities. The M&E framework provides outcomes, outputs, indicators, baselines, targets, responsible agencies and frequency of reporting for different thematic sectors including GESI. Both qualitative and quantitative indicators are included. They include participation of women and excluded groups in annual planning processes, subject committee meetings, public hearings and social audits. Targets are disaggregated by sex, caste and ethnicity.

In 2012, a study by ADDCN and UNICEF highlighted that GRB classifications of the MoFALD budget, as in other sectors, were inconsistent over time (ADDCN 2012). It also indicated fluctuations in the percentage of budget allocations for directly gender responsive, indirectly gender responsive and neutral programmes. Although MoFALD formulated its own GRB guidelines in 2009, these are not used locally because of limited awareness among local planning stakeholders and their poor commitment to gender balance. To deal with this problem, MoFALD agreed that LBs should allocate a mandatory 10 % of central capital grants to women's programmes.

The study identified several challenges in the application of GRB methodology. These include inaccurate understanding of GRB, inadequate administrative infrastructure, coordination problems and information gaps. Elaborating on information gaps, the study highlighted that LBs and sectoral agencies were often ignorant of local development guidelines and GRB. The study recommended improvement in women's representation and simultaneous capacity building for all stakeholders in planning and GRB at the local level.

On the other hand, an assessment by the UN in 2012 of the utilization of targeted capital grants pointed out that all LBs are expressing their commitment towards inclusive development at the local level through decisions made by district/village and municipal councils. It noted that the number of projects implemented through targeted grants and local resources did increase. Of the total projects implemented, 24 % was meant for the targeted population in 2008–2009; this increased to 28 % in 2010–2011. However, although the participation of targeted groups in planning and implementation was increasing, it was not yet satisfactory as the study reported that only a limited number of women (22.2 %) and DAGs (25.9 %) were actively involved in decision-making processes in VDCs. The study recommended orienting all the stakeholders on targeted fund utilization and GRB to ensure ownership, transparency and accountability of LBs. It also recommended capacity-building sessions for officials serving at different levels in LBs, including facilitators and motivators, so that they could formulate and implement programmes more effectively. A successful and not-so-successful case study are described below.

Box 9.3: Case Study 3: Women's Economic Empowerment

Jayalaxmi Women's Group with 35 women members wanted to undertake socio-economic activities at the community level. In FY 2010/11, group chairperson Faguni Tharu learnt from the chief of District Agriculture Office about a proposal to grow elephant foot yam (*Wole* in Nepali) as an innovative income-generating activity for women. With the financial support of Ramgram Municipality and technical backstopping of DAO, they started *Wole* farming at 35 locations from June 2011. This product has health benefits and religious significance, and therefore is expected to fetch a good price with consistent high demand in the domestic and nearby markets (ADDCN 2012).

Box 9.4: Case Study 4: Endless Investment

Dip Jyoti Women Cooperative Ltd was established in 2012 with 130 Janajati and Dalit women. A cooperative building construction activity was started in 2008/09 and as of now about Rs 300,000 has been spent through the capital grants allocated by the DDC and the VDC for three consecutive years. The amount due from the VDC as block grant for the last year was yet to be released. The women's group members have spent from their pockets and borrowed Rs 50,000 from a school nearby for settling urgent claims. The new building is roofed, but half of the doors and windows are yet to be fixed. Due to heat and rain over the years, doors and windows made from local soft wood have begun rotting. It is not certain when the building will be complete and there is no plan as to how the new building will be utilized for the socio-economic empowerment of the local women from disadvantaged groups (*Source* ADDCN 2012).

A study conducted in early 2013 by UN Women to assess local funding of women's economic empowerment and reduction of GBV at the local level pointed out that LBs spend an average 10 % of earmarked funds on projects directly benefiting women. It was 12.18 and 9.4 % in DDCs, 7.7 and 12.7 % in municipalities and 9.13 and 9.53 % in VDCs in 2012–2013 and 2013–2014, respectively, against a 10 % provision in policy. The rest of the money was spent on infrastructure development, social and capacity development, skills and institutional development activities. Infrastructure and trainings were not only a more visible legacy, they were also easier to deliver than social goals. The study team found women's participation to be about 30 % in different committees at the district level and about 55 % at the VDC level. This was because the DDCs include official representatives of sector line agencies and representatives of political parties at the district level (all of which comprise mostly men) while VDCs have mostly direct representation from citizens, allowing the statutory provision to be implemented.

Despite the successful implementation of GRB programmes and projects, significant challenges remain in the process of institutionalizing GRB. Some of them are discussed below. Some sectoral policies (related to infrastructure, public financial management, etc.) that directly influence local development initiatives are often gender blind, discriminatory, ineffective and lacking transparency. Lack of technical capacity in gender as well as participatory development means that many policies are not implemented effectively. Often, not just government officials, but NGOs hired to implement programmes lack expertise. Gender staff are often overworked and lack adequate administrative infrastructure.

As Case Study 2 shows, women's involvement at various levels is sometimes only a formality, and it is not effective in ensuring that women's voices are heard and their needs met. Women's participation in decision making, though mandated, is either not implemented or not effective when implemented. Deeply ingrained cultural norms and practices continue to undermine progress in gender equality as women are primarily viewed as wives and mothers. Many local leaders, both women and men, are patriarchal and insensitive to women's strategic needs. Men continue to offer strong resistance to policy provisions that target greater gender equality in the community and the family.

Usually, women and marginalized groups are busy with household and livelihood activities. They often reach meetings late, which limits their understanding of the subject matter and situations. Even though they are present as per the attendance records, their needs and interests remain neglected and unaddressed in these meetings. Moreover women's voices are generally less effective to influence the decisions as they are shy and neglected (based on discussion with WCF women members). The situation can be transformed with strong procedures for mandatory recording of all women's voices before decision making. Discussions in the field indicate that gender is often equated with activities for girls and women rather than activities that address the unequal relationship between women and men. Many government officials interpret gender integration as a focus on women rather than as a process of carrying out a gender analysis and identifying gender inequalities, whether for women or men, that can be redressed through appropriate programmes.

Other barriers to women's participation include lack of resources to pay for transportation, restrictions on women's mobility, time constraints, lack of education and knowledge of rights, lack of formal training in political processes, political interference and corruption. Women, particularly poor women, generally lack time and money to travel to various meetings due to their heavy productive and reproductive workload. To participate effectively in local decision-making institutions, they need to abide by complicated regulations which are often not simplified or translated into local dialects.

GRB implementation involves sequential steps during the different stages of the budget cycle: analysing the status of gender inclusion, examining the gender responsiveness of policies/programmes/projects, assessing budget allocations, identifying gaps, estimating expenditure required to bridge gaps and reallocating budgets, tracking actual spending and decision making and assessing gender disaggregated outcomes and impacts. In Nepal, current GRB initiatives do not systematically follow these steps. An effective GRB system must encompass the complete budget cycle and include detailed assessments of each programme in the above sequence.

9.9 The Way Forward

Nepal's GRB currently encompasses input-based assessment of benefits to women and their empowerment, with a focus on inputs and project targets rather than on measurable results. The methodology needs to link the GRB indicators of each sector with the institutional outputs and indicators of that sector, leaving out those that are irrelevant and adding those that are relevant. Additional requirements for the successful institutionalization of GRB across Nepal are: improving women's representation in decision making, enhancing stakeholder capacity in planning and GRB, reinforcing GRB classification at all levels, adopting MOF-GRB guidelines at district and central levels, making results-based monitoring systems more gender friendly, reviewing and revising GESI policy to reinforce institutionalization of GRB, and mainstreaming gender in policies and plans to redress inequalities and promote women's economic, social and political rights.

With these challenges in mind, Phase II of LGCDP (2013–2017) has been designed to focus on citizens' empowerment for better participation in local governance, especially the excluded and marginalized sections of society (Ministry of Local Development 2013). LGCDP II provides an overall framework for strengthening decentralization, devolution and improved local governance for the effective delivery of basic services and the empowerment of citizens, especially women, children and DAGs and their institutions. It has been designed as a framework programme with nine output and four outcome areas. One of the outcomes focuses on empowerment of citizens, especially women, children and DAGs, and their institutions. The aim is to help them meaningfully participate in local planning, budgeting, monitoring and decision-making processes and access services available at the local level. LGCDP II has four significant cross-cutting themes including gender equality and social inclusion. The programme targets GRB implementation in at least 50 DDCs and 40 municipalities by the end of 2016–2017. It encompasses all the programmes, actors and institutions in the local governance process through enhanced networking, collaboration and coordination mechanisms and systems to achieve its outputs and outcomes.

With this framework, the LGCDP II will hopefully provide the systems, procedures, structures, tools and capacities to facilitate the transformation of LBs into LGs capable of implementing a decentralized and effective local governance system, particularly in the context of GRB, women's rights and empowerment. In such a context, gender and participatory budgeting can better complement and strengthen each other.

References

Association of District Development Committees of Nepal (ADDCN), 2012: "A Study Report on the Assessment on the Utilizations of Capital Block Grants and Internal Sources for Children, Women and Disadvantaged Groups through Local Bodies in Nepal" (ADDCN with the help of UNICEF: Nepal).

Ministry of Finance, 2012: *Gender Responsive Budgeting Guideline* (Kathmandu: Government of Nepal).

Ministry of Local Development, 2008: *Local Governance and Community Development Programme (LGCDP)—Programme Document* (Kathmandu: Government of Nepal).

Ministry of Local Development, 2013: *Local Governance and Community Development Programme (LGCDP) Phase II—Programme Document* (Kathmandu: Government of Nepal).

Sharp, Rhonda, 2003: *Budgeting for Equity: Gender Budget Initiatives within a Framework of Performance Oriented Budgeting* (New York: UNIFEM).

The Local Governance and Community Development Program in Nepal (LGCDP), 2012: "Mainstreaming Gender Equality and Social Inclusion" (with the help of LGCDP Joint DPs: Nepal).

Chapter 10
Gender Responsive Budgeting: State and Civil Society Initiatives at National and Sub-national Levels in Indonesia

Agus Salim

Abstract The Indonesian government began the process of institutionalizing gender concerns with the presidential instruction on gender mainstreaming in 2000. A regulatory framework governs the GRB initiative at the national and sub-national levels including both government and civil society actors. Over the past decade, GRB initiatives have been conducted by both government and civil society at both national and sub-national levels. Cooperation between civil society and government, while not without its tensions, has been critical to the promotion of gender responsive planning and budgeting in Indonesia.

Keywords CSO · Sub-national level · National strategy · Gender mainstreaming · Gender budgeting

10.1 Introduction

Since 2009, Indonesia has been institutionalizing Gender Responsive Budgeting (GRB) at the national and sub-national level, where it is known as Gender Responsive Planning and Budgeting (GRPB).[1] This strategy emerged from the Indonesian government's commitment to gender equality and justice, marked by the formation of a ministry in 1978, now called the Ministry of Women's Empowerment and Child Protection (MOWECP), specifically to promote these goals.

This chapter traces the Indonesian government's process of institutionalization of gender concerns beginning with the presidential instruction on gender mainstreaming in 2000. First the regulatory framework that governed the GRB initiative is discussed, followed by GRB initiatives at the national and sub-national levels by both government and civil society actors. This is followed by a discussion and evaluation of GRB initiatives conducted by both government and civil society at these two levels. Though

Agus Salim, Programme Manager, Pattiro, Jakarta, Indonesia. Email: agus@pattiro.org.

[1]GRPB is a gender responsive planning and budgeting system that has GRB as an output.

C. Ng (ed.), *Gender Responsive and Participatory Budgeting*,
SpringerBriefs in Environment, Security, Development and Peace 22,
DOI 10.1007/978-3-319-24496-9_10

the relationship between the two sectors has not always been smooth, the synergy and cooperation between civil society organizations (CSOs) and government has been critical to the promotion of gender responsive planning and budgeting in the country. After discussing the challenges and opportunities of this collaboration, the chapter ends with a brief overview of the GRPB plan for the future.

10.2 Gender Mainstreaming and GRB Regulatory Framework

The Indonesian government's commitment to gender justice and equality led to the issuance of a gender mainstreaming policy to be applied to all development programmes through Presidential Instruction No. 9/2000 on Gender Mainstreaming in National Development and other regulations as seen in Fig. 10.1. Gender mainstreaming became the strategy to integrate gender perspectives into planning, budgeting, implementing, monitoring and evaluating policy and development programmes. Thereafter, planning and evaluating policy and gender sensitive development programmes according to the roles and functions within the scope of national and sub-national governments gained importance.

Gender mainstreaming was one of the operational foundations for implementing a development agenda and therefore, gender mainstreaming principles became reflected in development policy. It was expected that gender integration in planning and budgeting in accordance with these principles would create a more effective, accountable and fair allocation of human resources. Therefore, the gender mainstreaming policy was incorporated into the planning and budgeting system as GRB in 2010 through the 2010–2014 National Medium-Term Development Plan (RPJMN). Various ministries, institutions and relevant stakeholders were then required to develop policies, indicators and gender-based targets.

10.3 National Level GRB Initiatives

As noted earlier, Indonesia's GRB initiative began with the implementation of Presidential Instruction No. 9/2000. A Gender Mainstreaming Working Group (*Pokja* PUG) was formed in each Ministry/Institution to increase stakeholder awareness of gender mainstreaming through various public events and training. The scope of gender mainstreaming in the Instruction covered: (1) planning, including gender responsive planning and budgeting, (2) implementation and (3) monitoring and evaluation.

As of 2004, 38 gender responsive programmes were implemented in various development sectors such as education, agriculture, labour, law, social welfare, health, family planning and environment. After the policy was issued, the National Planning and Development Agency (Bappenas), as one of the main drivers of GRB implementation, conducted research on gender mainstreaming implementation. In

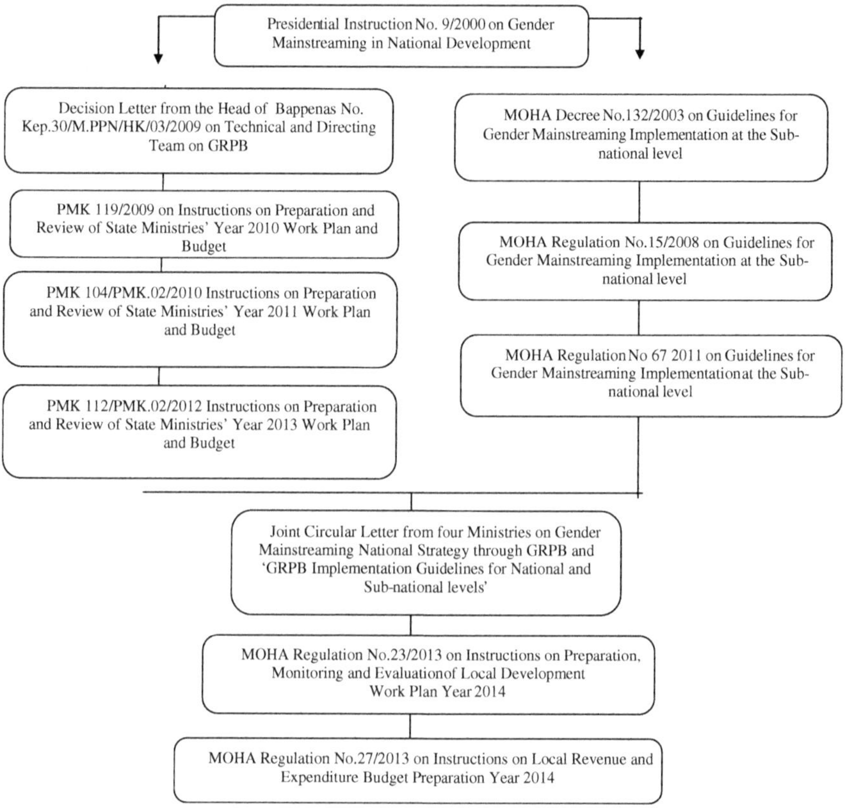

Fig. 10.1 Gender mainstreaming policies in Indonesia. *Source* The author

2005, results showed that GRB implementation was satisfactory although a few improvements were needed. Recommendations included the need for: a legal basis for gender mainstreaming implementation, strengthening stakeholders' understanding of gender mainstreaming as a basic principle in development and forming a special implementation unit such as a gender mainstreaming working group.

Again, in 2007, Bappenas, as the coordinating agency for the national development plan, conducted 'Gender Analysis in Development' research to evaluate gender mainstreaming efforts since the enactment of Presidential Instruction No. 9/2000. This was conducted in 18 ministries/institutions, seven selected provinces and seven regencies/cities. However, this time, it showed that commitment from the ministries, provincial and city governments in implementing gender mainstreaming was not yet optimal. More concrete, non-abstract gender mainstreaming content was needed that could solve related issues, provide a better gender perspective, and also integrate this perspective into planning and budgeting.

Implementation of the GRPB plan began with the issuance of the Ministry of Planning's (Bappenas) Decree No. 30/M.PPN/HK/03/2009 on Technical and Directing Team for GRPB. The team's task was to coordinate cross-ministerial and cross-sectoral GRPB implementation. It was led by the Head of Population, Women's Empowerment and Child Protection Directorate at Bappenas. A national GRB workshop was held in 2009 and international GRB experts such as Debbie Budlender were invited. Budlender discussed the GRB experiences of other countries and introduced the Gender Budget Statement (GBS) as a tool to realize the government's commitment to implement GRB. Following the workshop, three Ministries emerged as the main drivers of GRPB implementation at the central level: Bappenas, Ministry of Finance and MOWECP. The GRPB implementation regulation was then integrated into the regulation of budget preparation issued annually by the Ministry of Finance. Thus, the gender mainstreaming policy was incorporated into the planning and budgeting system as GRB in 2010.

At the same time, the government and civil society collaborated to draft the National Strategy (Stranas PPRG) on the Acceleration of Gender Mainstreaming through GRPB. The draft was completed at the end of 2012 and was endorsed by Bappenas, the Ministry of Finance, the Ministry of Home Affairs (MOHA) and the MOWECP. In 2012, these four ministries driving GRB issued a Gender Mainstreaming National Strategy through GRPB to accelerate gender mainstreaming implementation which was aligned with RPJMN 2010–2014 to support good governance and to meet the goals of the Millennium Development Goals (MDGs). This strategy aimed to ensure that gender mainstreaming implementation in the development cycle was better organized, systematic and coordinated at the national and sub-national level.

Stranas PPRG included National Implementing Guidelines for Gender Responsive Planning and Budgeting for ministries/institutions and local governments. It was coordinated by the MOWECP and consisted of technical guidelines to integrate gender issues into planning and budgeting documents. The interesting aspect related to this national strategy and its implementing guideline was that they were collectively developed with international organizations and civil society as development partners.

The National Strategy consisted of a general and a special strategy. The general strategy focused on strengthening the legal basis and coordination between the driving institutions and/or between the driving and implementing institutions. The special strategy elaborated on the general strategy. GRPB implementation at the sub-national level needed a stronger legal basis and at the national level, annual GRPB implementation targets needed to be confirmed. Stronger coordination was needed through a series of steps that covered planning, budgeting, implementation, monitoring and evaluation. MOHA attempted to strengthen GRPB by creating a National Joint Secretariat to facilitate the implementation of the Gender Mainstreaming National Strategy through GRPB at the local level. At the same time, MOHA sent a circular to all Governors in Indonesia to establish a Local GRPB Secretariat at the provincial level.[2]

[2]The GRPB Secretariat focuses on ensuring that results are achieved and documented. The Gender Mainstreaming Working Group facilitates and strengthens Gender Mainstreaming and offers technical support to refine the results achieved.

Simultaneously, MOWECP, as the main facilitator, with The Asia Foundation and PATTIRO,[3] started promoting GRPB through socialization, training and assistance to various regions in Indonesia. Through a series of decrees issued by the different ministries, including the Ministry of Finance, MOHA and others, the government offered the legal basis for a GRPB pilot which was implemented in seven ministries/institutions with the three Ministries mentioned above driving the process while four other Ministries (Health, Agriculture, Education and Culture as well as Public Works) acted as implementing bodies. The Decrees also provided guidelines for regions in preparing gender integration strategies through planning, budgeting, monitoring and evaluating programmes and development activities.

As noted earlier, gender mainstreaming was integrated into the planning and budgeting system of the 2010–2014 National Medium-Term Development Plan (RPJMN). Specifically, gender mainstreaming through GRPB was included in the development policy strategy to:

1. harmonize regulations and their implementation at all government levels by including all stakeholders to improve the quality of lives and women's empowerment;
2. conduct prevention, provision and improvement efforts to protect women against various forms of violence; and
3. increase gender mainstreaming and institutional capabilities. (Population Directorate et al., undated)

Accordingly, the 2010–2014 RPJMN goals to improve gender equality were to (1) improve the quality of life for women especially in the health, education and economic sectors including access to politics and control of resources; (2) increase the percentage of women victims of violence who have access to systems for filing complaints; and (3) increase institutional effectiveness in planning, budgeting, implementing, monitoring and evaluating gender responsive policies and development programmes at the national and sub-national levels.

Gender mainstreaming was then also included in the Long-Term National Development Plan (RPJPN) 2005–2025 with emphasis on women's empowerment through improving quality of life and role of women in various development sectors; decreasing violence, discrimination against and exploitation of women; and strengthening institutionalization and gender mainstreaming network.

The GRB movement has been in existence for 13 years in Indonesia and various agencies have been evaluated throughout this period. The latest study from 2009 to 2011 by Bappenas and MOWECP—the two main drivers—depict the evaluation results in GRPB implementation of seven ministries as shown in Table 10.1.

[3]PATTIRO (*Pusat Telaah & Informasi Regional*) was established on 17 April 1999 with the aim of social justice and fulfilment of citizens' basic rights through good governance and public participation in Indonesia, particularly at the local level. PATTIRO has worked in nine provinces on various themes such as access to public services (education and health) for the poor, gender budgeting, community participation, information transparency, voter education, development planning and transparency of oil revenues.

Table 10.1 Evaluation result in ministries/institutions

No	Ministry/institution	2010		2011	
		Programmes	GRB 2010	Programmes	GRB 2011
1	Public Works	8	1.518	8	91.105
2	Agriculture	–	0	3	19.818
3	Health	8	4.908	6	11.356
4	Education	1	0	1	550
5	Finance	3	700	5	1.118
6	Women's Empowerment	3	1.840	2	1.371
7	Development Planning	–	0	1	135
Total		23	8.966	26	125.453

Source Directorate of Population, Women's Empowerment, and Child Protection, Bappenas (2011)
Note GRB (in US$ million)

GRB implementation in these seven ministries/institutions for FY 2009 and 2010 reflects government seriousness in implementing GRPB, although the commitment of some ministries or institutions seems weak. The total budget for gender responsive programmes has increased significantly from $8.966 million to $125.453 million. This increase was obtained due to a commitment to undertake activities and analyse programmes which led to increased gender responsive allocation from the Ministry of Finance. The Ministry of Public Works conducted gender responsive analysis for eight of its programmes in 2010 and 2011 and showed a sharp increase in budget allocation.

The Ministry of Finance's GRB recapitulation data drafted by ministries/institutions in 2012 and 2013 showed their commitment to implement GRB into their programmes. In 2012 and 2013, 19 ministries implemented GRB, including the Supreme Court, the Ministry of Justice and Human Rights, Ministry of Finance, Ministry of Agriculture, Ministry of Industry, Ministry of Energy and Mineral Resources, Ministry of Education and Culture, and many others.[4]

As an incentive, MOWECP as the driving ministry for gender mainstreaming has been conferring an *Anugerah Parahita Ekapraya* (APE)[5] award since 2004 to ministries or institutions and regions that have implemented GRPB. The Ministry of Public Works and Ministry of National Education (now called the Ministry of Education and Culture), the Central Java province and the East Java province have been repeat winners of the APE awards.

[4]Explanation provided by Vice Minister of Finance when launching National Strategy, 2013.

[5]The *Anugerah Parahita Ekapraya* is awarded to a ministry, organization or local government (both provincial and district/city) that has committed to and implemented strategies on gender mainstreaming, empowerment of women and protection of children in various development sectors.

GRPB implementation is unique in Indonesia because it is conducted simultaneously at the national and sub-national levels and implemented within the government or by civil society. Government support comes in the form of pro-GRPB government policies and civil society contributes by pushing for the implementation of such policies at the national and sub-national levels. Since GRPB implementation, CSOs in Indonesia have been actively advocating GRB implementation among the public and the government. Focus group discussions (FGDs) conducted by CSOs pinpointed issues faced by women's groups and the poor. These groups are marginalized and only allocated a small proportion of the budget, and hence, CSOs are looking for solutions that can be accommodated in the local development work plans and expenditure budgets. Within governmental agencies, CSOs are also strengthening the institutionalization of gender mainstreaming through training and technical assistance.

10.4 GRB Initiatives at Sub-national Level

Stranas PPRG was discussed by Bappenas, MOWECP and the Ministry of Finance before being presented to the MOHA since no MOHA Regulations (*Permendagri*) were available on GRPB implementation at the sub-national level. Since then the Annual Policy issued by the MOHA (on Preparation, Monitoring and Evaluation of Local Development Work Plan 2014 and Instruction for Preparation of Local Revenue and Expenditure Budget 2014) has mandated that sub-national governments consider Stranas PPRG when preparing their Local Development Work Plan and Local Revenue and Expenditure Budget (known as *Anggaran Pendapatan Belanja Daerah* or APBD).

Sub-national autonomy was a key part of the 1998 democratic reform process in Indonesia and it was expected that the democratic process concretized through sub-national autonomy would not stop at the policy level but would include all elements of society as targets of and actors in development. To achieve this, civil society and community participatory spaces were opened in the development planning and implementation process. This, in turn, increased expectations of good governance. Local government units and civil society organisations thus became key players in GRPB implementation.

To accelerate the institutionalization of gender mainstreaming at all Local Government Working Units (SKPD) in the provinces and regencies/cities, MOHA asked local governments to form Gender Mainstreaming Units (*Pokja* PUG) and Focal Points. Local government leaders, through Governor/Regent/Mayor Decrees, were to form *Pokja* PUG and assign the Head of the Local Development Agency (Bappeda) to lead it. The Head of the SKPD working on women's empowerment issues was to be the Secretary of the *Pokja* PUG, and other Heads of SKPD were to be members.

The gender mainstreaming focal point in each SKPD, either in provinces or regencies/cities, would consist of officials working on plans or programmes. It would be helmed by the Head of SKPD in each unit. The focal point is supposed to understand and appreciate the importance of sex disaggregated data for gender analysis while drafting a gender responsive development profile. In general, planning and budgeting in each SKPD was to take note of the gender gaps in development. Moreover, in formulating the SKPD Work Plan and SKPD Work Plan Budget, focal points' tasks were to conduct gender analysis by using the Gender Analysis Pathway (GAP) instrument and GBS to direct activity targets and materials for greater gender justice and equality in each SKPD.

Some duties of Provincial and Regent/City *Pokja* PUG as listed in the MOHA Regulations include promoting and facilitating gender mainstreaming for each SKPD; conducting gender mainstreaming socialization and advocacy with leaders at the sub-district, urban village and village levels; preparing the annual work plan; promoting GRPB implementation; reporting to the Mayor/District Head through the Vice-Mayor and Vice-District Head; formulating policy recommendations for the Mayor or Regent; drafting gender profiles; preparing a Local Action Plan (RANDA) for gender mainstreaming and monitoring its implementation in each institution; forming a technical team to analyse the local budget; and promoting the election and implementation of a Focal Point in each SKPD. The SKPD Focal Point at provincial and regency/city level is supposed to promote gender mainstreaming and gender analysis of policies, programmes, and activities in each working unit; facilitate drafting sex-disaggregated data and a gender responsive work plan and budget preparation at the SKPD; conduct training, socialization, and advocacy of gender mainstreaming for all SKPD employees; and report on gender mainstreaming implementation to the Head of the SKPD.

Since these policy directions, efforts have been made at the provincial level to implement GRPB in various programmes. The results of an evaluation of GRPB pilot implementation, conducted by Bappenas, in four provinces in 2011 are shown in Table 10.2.

Table 10.2 shows that the ministries/institutions and SKPD at the provincial level have at least one GBS document. However, most of the activities proposed in GBS are not prioritized, which minimizes impact. Further, the legality of GRPB implementation is unclear and GRPB is still isolated from local planning priorities because it uses the ad hoc line through *Pokja* PUG. Commitment of the policy makers or GRPB driver level is weak as are human resources for GRB implementation.

Table 10.2 Evaluation result in four provinces

No.	Province	Sub-national government units	Total GRB
1	Banten	9	2.316
2	D.I. Yogyakarta	9	1.063
3	Jawa Tengah	16	24.005
4	Jawa Timur	9	5.878
Total		43	33.262

Source Directorate of Population, Women's Empowerment, and Child Protection, Bappenas (2011)
Note GRB (in US$ million)

As democratic and participatory spaces widen as the result of the government's efforts to implement good governance practices, civil society movements which can better convey people's aspirations are also expanding. CSOs play their role by handling basic issues such as providing filtered data, analysing gender issues and providing technical assistance to appropriate sectors. Previously, the role of facilitator was undertaken by the government through its gender mainstreaming working units, and focal points or the GRPB technical team. But some of these functions are now undertaken by CSOs.

Today, socialization, training and technical assistance in strengthening gender mainstreaming are conducted continuously and have become joint agendas between national level government and CSOs for accelerating GRPB implementation at the sub-national level.

10.5 Role of CSOs

CSOs in Indonesia have been lobbying for a long time for community interests, representing various communities at the local level in pushing for improvements in public service and increased budget allocations. This has helped bridge the gap between communities and local governments although community understanding of public policy and budget management still needs to be improved. At the local government level, technical assistance from CSOs provides important feedback on issues or policies that have been targeted for change.

In several regions, civil society has lobbied for the formation of GRPB institutions such as the *Pokja* PUG and the Focal Points. CSOs have assisted planners at SKPD and developed tools for analysis and data filtering. Indeed, civil society's role and involvement has changed: earlier, political advocacy was undertaken by pushing transparency in budget reallocation; now, however, the role may be re-defined as critical engagement and even building the capacity of governmental institutions in such processes. CSOs have helped strengthen institutions to promote GRPB policies and have become an integral part of the acceleration of GRPB implementation in the region. Support from international donors and organizations such as The Asia Foundation, UN Women, *Deutsche Gesellschaft für Internationale Zusammenarbeit* (GIZ) and others has had an impact on the increasing number of CSOs working on and advocating GRB issues.

Strategies utilized by CSOs to promote GRB implementation include: (1) partnership—providing technical assistance to national or sub-national governments; (2) facilitation—encouraging GRPB policies at the national or sub-national level; and (3) advocacy for the current policy—through State institutions that are directly responsible for implementing the policy either at the national or sub-national level. At the same time, CSOs also maintain a critical view on various planning and budgeting documents at both national and sub-national level.

Since 2001, PATTIRO, with support from The Asia Foundation, has been strengthening women's participation in public policy in Tangerang (Banten

Province), Semarang and Surakarta (Central Java Province). The programme was followed in 2003 by efforts to promote women's participation in sub-national budgeting. The aim was to facilitate and strengthen the involvement of women and women's groups in all public participatory spaces. This then transformed progressively into a gender budget advocacy movement with women's groups, especially in the Central Java region.

PATTIRO's pilot programmes in Semarang, Surakarta and Tangerang focused on raising awareness in the community, highlighting community needs and strategizing for budget reallocation at the sub-national level. Through the programme, local government budgets incorporated several needs expressed by the communities including road repairs in Podhorejo, additional budget for oil fuel during flooding, provision of clean water infrastructure in Panggung Kidul (all in Semarang, Central Java Province) and funds to form cooperatives in the Karang Timur in Tangerang. In Semarang, the budget was reallocated to procure waste collection bicycles that would benefit the community instead of procuring a vehicle for the sub-district head. This was due to people's participation in the process of decision making, rather than a top-down decision about use of funds by government officials. Further, though waste was everybody's problem, women were more focused on ensuring efficient waste collection due to their gender roles in the family.

PATTIRO played a key role in providing better understanding of the technocratic process of planning and budgeting and encouraged society's involvement through the Society Development Planning Forum (*Musrenbang*) that was conducted from the village to the regency levels. Active citizens' participation improved society's understanding of the government's development planning process through needs assessments, sub-national budget document analysis and mapping of political actors.

In 2008, PATTIRO actively supported the formation of Community Centres (CC) in various sub-districts in Lebak regency, Banten Province, through the Participatory Budgeting Expenditure Tracking (PBET) programme. These centres are actively engaged in development budgeting and planning undertaken by government, from the village to the regency level. Community engagement in this development planning process generated great enthusiasm for involvement in these planning forums, particularly in areas that are yet to benefit from development. Various proposals have been delivered in these forums by the centres such as village irrigation, small business assistance and support to village midwives. CCs also met with DPRD (*Dewan Perwakilan Rakyat Daerah*/District House of Representatives) or local parliament members to support proposals to budget documents during discussions on the Budgetary General Policy (*Kebijakan Umum Anggaran* or KUA). This participatory process helps strengthen CC proposals through representations in parliament. The beginning of the mentoring process is challenging, as it involves developing community group awareness and inviting community members to play an active role in the development and planning process and championing their needs in parliament.

The main challenges to government-CSO relations include trust because government perceives CSOs as being critical of its policies and as such officials worry that documents provided may be misused. However, Indonesia has already passed

Law No. 14/2008, on Public Information Openness, which provides guarantees to the community that is requesting information, including guarantees to public bodies when information is misused.

10.5.1 Achievements by Civil Society at Sub-national Level

The GRPB idea comes from the concept that the government is not the lone actor deciding public policies. Good governance assumes that the government must also allow elements of civil society to voice their concerns, thus opening up planning and budgeting information channels to society. CSOs generally conduct deeper analysis and are more responsive to marginalized groups such as women and the poor. This implies that non-governmental actors such as CSOs play an important part and can contribute to the management of public policy. Communities, together with CSOs, have pushed for government commitment to fulfil the needs of women's groups through budgetary allocations for programmes. For example attempts have been made to focus on the actual needs of communities with regard to better services for women, such as better maternal and child health programmes, literacy programmes and other gender equality programmes.

The above efforts to implement GRB have created an empowered community, especially among women who are vocal about their views on village development and participate in village planning and budgeting forums. At the same time, budget reallocation has been a key strategy in civil society's approach in promoting GRPB implementation. CSOs have demanded the reallocation of the government budget after their analysis of government programmes and activities found them ineffective, inefficient and unequal. As a result, budget allocations increased in some instances and were cut where necessary, as articulated in the cases below.

Box 10.1: Case study 1: Fuel assistance to cope with tidal floods. Source: PATTIRO Report (2006).

The CSO PATTIRO concentrated on mentoring women in Semarang, Surakarta/Solo and Tangerang. It undertook needs assessments through FGDs that delved into the problems faced by women and searched for solutions. In Semarang, FGDs with women uncovered various issues including health, flooding and educational (high fees) problems. Discussion forums concluded that a fuel assistance programme for tidal vacuum pumps would be proposed to deal with the recurring problem of tidal flooding. Tidal floods add to women's burdens as after the tidal waters have receded women's work (cleaning of homes) increases. Mothers must care for children who suffer from skin diseases following exposure to the flood waters. The fuel assistance programme for tidal vacuum pumps would help the community, particularly women in Panggung Kidul village, and reduce their burden. Furthermore, the community undertook

APBD analysis to find sources of inequality and wasteful spending. The results led to the struggle for this community's proposal with the relevant data, analysis and needs to be submitted to the local government and the DPRD.

Box 10.2: Case study 2: More funds to fight malnutrition. Source: PATTIRO Report (2008).

In Tasikmalaya *Kabupaten*, according to 2007 data, children suffering from malnutrition totalled 600 with 32 cases of maternal mortality reported. This issue became the advocacy focus for PATTIRO, because malnutrition mirrored the marginalization of infants and the limited governmental attention paid to the fulfilment of this need. It multiplied the burdens faced by women who were traditionally responsible for nurturing and caring for children. Overcoming malnutrition meant accommodating the needs of infants and fulfilling practical gender needs.

Analysis of Tasikmalaya *Kabupaten* APBD showed that the local government was not committed to resolving these issues. This was evident in the 2008 APBD policies, specifically the budget allocation for handling malnutrition and maternal mortality. In the 2008 APBD, only Rp. 119,000,000 was allocated for handling malnutrition. This was extremely dissatisfactory when compared to the number of children suffering from malnutrition, as it provided for only Rp. 48,745 per child. According to a doctor who worked in a community health centre (*puskesmas*), the minimum intervention needed for treating malnutrition was Rp. 1 million per person. This allocation also fell significantly short when compared to the budget for coordination activities with central and other governments of Rp. 2,186,340,000 (this equates to the *bupati* and his deputy spending Rp. 182,195,000 every month just for coordination). As a result, women's groups held audiences with stakeholders, disseminated the results of their studies and requested that the government reallocate some of the coordination budget for overcoming malnutrition.

Similarly, in Surakarta/Solo city, Central Java Province, the FKKP (Healthcare Cadre Communication Forum) succeeded in requesting the city government to increase its healthcare budget for children and the elderly. Money allocated for the children's healthcare centre rose from Rp. 800,000 to Rp. 1,800,000/month to Rp. 1,440,000/month for the elderly healthcare centre.

Over the years, civil society which began with grassroots awareness raising has expanded its role to policy promotion and gender responsive programme recommendation at the local and national levels. It also encourages the government to implement gender responsive programmes and integrate them at programme planning and budget policy levels. The community has succeeded in pushing for Local

Regulations (*Perda*) to be implemented in several regions, such as the Local Regulation on Public Service in Jeneponto in 2007, Local Regulation on Educational System No. 18/2006 in Gresik, Mayoral Regulation on Regional Health Insurance in 2009 in Pekalongan and Operational Manual to Accelerate Development of Family Welfare in Pekalongan (PATTIRO 2005).

10.6 Government-CSO Collaboration

10.6.1 Opportunities

Institutionalization of GRB in Indonesia is an on-going process due to strong political will of the main government agencies in driving GRB implementation at the national and sub-national levels. In addition, collaboration between the Indonesian government and civil society, at both levels, has strengthened the further implementation of GRB.

The legal basis for GRB was provided by laws at the national level followed by decrees at the local level laying down guidelines for implementation of GRPB. Political commitment from the driving ministries has been found fruitful, if not necessary, in accelerating GRPB implementation, as attested by these decrees which were issued by MOHA after the National Strategy was implemented. For example, MOHA requested all Governors to form GRPB secretariats that drives GRPB drivers at the sub-national level.

In Indonesia, civil society's work in analysing budgets has been significantly aided by the regulatory framework put in place by the enactment of Law No. 14/2008. The law gives CSOs and communities access to information related to planning and budgeting held by the government. Greater transparency on policy documents issued by the government and greater accountability is also encouraged. This then creates better governance in Indonesia. Through this reform, communities and CSOs attempt to integrate gender issues systematically in the planning and budgeting processes. Gender integration was achieved through multiple steps taken by CSOs in promoting GRB including preparing sex-disaggregated data and gender information systems; integrating gender perspectives in the Local Medium-Term Development Plan (RPJMD) which was then endorsed by Local Regulations (*Perda*); and institutionalizing gender mainstreaming through several phases, from provincial to SKPD level. It covers decisions from local leaders regarding *Pokja* PUG and decisions from heads of SKPD on focal points with their staff members.

Experiences from the Central Java Province, one of the most advanced in GRPB implementation, have shown that political commitment and the availability of capable staff play an important role in accelerating GRPB policy implementation. The existence of regulations and institutional support are also critical for GRPB acceleration. It has been proven that with technical support from civil society and political will from the top echelons of government, it is possible for the bureaucratic system to realize GRB in the national and sub-national SKPD Work Plan.

10.6.2 Challenges

There have been several challenges during the implementation of GRPB at the national and sub-national levels, which may be grouped into four categories: lack of technical capacity in GRB, high staff turnover, lack of commitment from the head of local agencies, and poor coordination between the different agencies. GRB capacity among government officials is a key concern because the paucity of capable facilitators and gender experts at SKPD to advocate for GRPB issues poses a serious challenge to integrating gender into different sectors at the local level. Gender Mainstreaming Working Groups have been established in almost all sectors. However, they are not optimal in undertaking their roles and tasks, including GRPB advocacy to unit heads, supporting the realization of GRPB, monitoring the implementation of gender mainstreaming in each agency and establishing technical teams to analyse local budgets. This has resulted in the inadequate implementation of GRPB in various areas.

Furthermore, the implementation of GRPB is sometimes constrained by the disposition of those at the helm. For example, meetings of the Working Group on Gender Mainstreaming always comprise staff who have no authority to decide on the matters discussed. Commitment by leaders is a variable that strongly influences the acceleration of GRPB policy implementation, as consolidation of resources, financial and political support and technocratic capacity require bureaucratic push. Poor coordination between ministries/institutions/SKPDs at the sub-national government has led to overlapping programmes being implemented. Despite all these challenges, the synergy between government and CSOs allows GRB implementation to gain strength each year.

10.7 Conclusion

Decentralization and democratization are two mainstream political changes in Indonesia as a result of the 1998 decentralization reforms. Democratization not only includes changes in the election system, but more importantly, it allows public participation at the local level, particularly in the public domain which affects people's everyday lives. Democracy aims to bring the State closer to the community and bring accountability, transparency, responsiveness, as well as community participation to local governments to improve social welfare. It allows citizens a direct channel to the government, as well as access and control over the government. Moreover, changes in governance often start from the local level since the local government has the authority to manage its administrations/jurisdictions.

Indonesia adheres to the top-down system of development planning, beginning with the national plan, regional, and finally, local development plan. GRB begins after the planning and budgeting documents are analysed through the Gender Analysis Pathway and the Gender Budget Statement instruments. However, in the process of its formulation, Indonesia also conducts a bottom-up system with the community through the multi-stage participatory deliberation forum.

Public service issues are likely to occur at sub-national and local levels. Therefore, CSOs focus on public service and policy advocacy and attempt to make changes locally which in turn can be a basis for policy change and betterment at the national level.

All these achievements notwithstanding, GRB in Indonesia needs further impetus. As outlined in the National Strategy, to accelerate GRPB implementation, integration between GRPB policies and related planning and budgeting policies need to be strengthened at the national or sub-national levels. This can be realized by prioritizing target GRPB issues in the 2015–2019 national medium term development plan, integrating GRPB into the local planning and budgeting policy (integration of GAP and GBS within the planning and budgeting format); strengthening the monitoring of GRPB implementation at the national or sub-national level; conducting continuous capacity building and technical assistance programmes for institutions implementing GRPB, creating gender champions/agents; and strengthening local CSOs working on and advocating GRPB. Similarly, CSOs are also encouraging the government to implement participatory budgeting.

One of the mechanisms to optimize GRPB implementation in the region is to maximize Sub-national APBD evaluation by integrating GRPB as part of the evaluation process. Regent/City APBDs are evaluated by Provincial governments and Provincial APBDs are evaluated by the MOHA. In this phase, the evaluating team would review the gender responsiveness of budgets to receive feedback in order to accelerate the implementation of the Joint Circular Letter on gender mainstreaming.

Finally, it is the joint efforts and collaboration between government and civil society in the context of the decentralization processes of democratic reform that will allow the people to have a genuine voice in the planning and budgeting of public funds to ensure that development provides fair benefits to all levels of society and to both women and men.

References

Directorate of Population, Women Empowerment, and Child Protection, 2011: *Evaluasi Terpadu Pelaksanaan Uji Coba Perencanaan dan Penganggaran yang Responsif Gender (PPRG) 2009–2010* [Integrated Evaluation of Pilot Gender Responsive Planning and Budgeting] (Jakarta: Bappenas).

PATTIRO, 2005: *Annual Report* (Jakarta: PATTIRO).

PATTIRO, 2006: *Annual Report* (Jakarta: PATTIRO).

PATTIRO, 2008: *Annual Report* (Jakarta: PATTIRO).

Penang Women's Development Corporation (PWDC)

The Penang Women's Development Corporation (PWDC) is a state agency which works towards the transformation of Penang into an advanced and progressive society through the promotion of equality among all women and men, girls and boys regardless of ethnicity, religion, gender or socio-economic background. PWDC believes that Penang's people are the key to her future and shared prosperity, and that all women and men, girls and boys must have the opportunities, rights and freedoms to develop their full potential as citizens who can contribute to achieving the visions and aspirations of the state.

Since its establishment in November 2011, PWDC has played a pivotal role as the key State Government agency working towards the mainstreaming of gender equality, social inclusion and good governance across all sectors in Penang. PWDC achieves this by working with partners to plan and implement gender responsive projects, raise public awareness of gender issues, and build capacity among individuals and communities. PWDC also advocates measures to bridge gender gaps between women and men in the economy, public sphere, politics and governance.

As Penang progresses towards its aspiration of being an advanced society, PWDC will continue to strategize and execute multi-level and multi-pronged approaches to bring about meaningful and lasting change. PWDC envisions a future in which all women and men, girls and boys will have the equal opportunity to fully contribute to and benefit from a Penang which values and upholds the principles of democratic participation, respect for diversity and individual dignity, and social inclusion and social justice.

C. Ng (ed.), *Gender Responsive and Participatory Budgeting*,
SpringerBriefs in Environment, Security, Development and Peace 22,
DOI 10.1007/978-3-319-24496-9

PWDC directors and staff who organised the GRB Conference. The permission was granted by PWDC who provided the photo

Vision

PWDC envisions a Penang which mainstreams gender into the policies and programmes of all sectors to achieve gender and social justice in the State.

Mission Statement

PWDC seeks to contribute to the transformation of Penang based on principles of substantive equality and good governance. Integral to our work is the recognition of women's diverse identities, women's representation in all spheres and the equitable redistribution of resources.

This will be achieved through research and advocacy on gender responsive policies, capacity development, strategic networking, and women's empowerment programmes; working in smart partnership with all levels of the Penang State and local governments, non-governmental and community-based organizations, the academe, the private sector and the public.

Objectives

PWDC has five key objectives, namely:

- To develop gender responsive policies and good governance;
- To promote awareness and understanding of gender and social justice;
- To enhance women's participation in the socio-economic and political life of society;
- To strengthen women's leadership and decision making in all sectors of society; and
- To enhance and promote an environment of inclusiveness within the multi-ethnic, multi-cultural and multi-religious context of Penang

PWDC Website at: http://www.pwdc.org.my
GRPB Project Website at: http://grb-pwdc.org.my

About the Editor

Dr. Cecilia Ng (Malaysia) is an independent researcher and women's rights advocate. She received her Master's from Harvard University and her Ph.D. in Sociology from Universiti Malaya. Her latest appointment was at the Women's Development Research Centre (KANITA), Universiti Sains Malaysia where she was a Visiting Professor (2007–2011) and Honorary Consultant (2011–2013). She was previously an Associate Professor at Universiti Putra Malaysia before she opted for early retirement in 2000. From 1994 till 1999 she was a Research Fellow at the United Nations University Institute for New Technologies in The Netherlands.

Upon retirement, she continued in the academe as Visiting Associate Professor at the Asian Institute of Technology, Thailand, where she was also an editor of the journal *Gender, Technology and Development* (2001–2006). In 2011, she was appointed as a Director, as well as the Advisor of the Gender Responsive and Participatory Budget Project, of the Penang Women's Development Corporation (PWDC), a state-run body established to promote gender equality and social justice in Penang.

She has conducted research and published widely on gender, development and work, with a focus on globalization, technological change, women's employment and gender and democracy. She is active in women's groups in Malaysia and has been involved in research, training and advocacy on gender issues since the early 1980s. Her recent publications include a co-edited volume on *Gender and the Digital Economy* (Sage, 2006), a co-authored book entitled *Feminism and the Women's Movement in Malaysia: An Unsung (R)evolution* (Routledge, 2006), a co-edited Special Issue (2010) on *Women in National Politics in Asia* in the journal

C. Ng (ed.), *Gender Responsive and Participatory Budgeting*,
SpringerBriefs in Environment, Security, Development and Peace 22,
DOI 10.1007/978-3-319-24496-9

Gender, Technology and Development and a co-edited volume (2011), published by Universiti Sains Malaysia Press entitled *Our Lived Realities: Reading Gender in Malaysia*.

Address: Dr. Cecilia Ng, 15-6-12 Desa Embun Emas, Lorong Sungai Kelian, Tanjung Bungah, 11200 Penang, Malaysia
Email: cncs77@gmail.com.

About the Authors

Giovanni Allegretti (Italy/Portugal) is an architect and Senior Researcher at the Centre of Social Studies, Faculty of Economics, Coimbra University (Portugal) where he coordinates the Ph.D. course "Democracy in the XXI century" and the Peoples' Observatory. He has conducted research on the participatory budgeting process in Europe and South America. He also worked as a trainer, evaluator and consultant on participatory budgeting in Asia and Africa, for different international organisations, universities and NGOs. He has published extensively on participatory planning and budgeting, in both national and international journals, in more than ten languages. He has been elected co-chair of the *Independent Authority for the Promotion of Participation* (APP) of Tuscany Region for 2014–2019.

Address: Arch. Giovanni Allegretti, Senior Researcher, CES (Centre for Social Studies), University of Coimbra—Room 23, Colegio S. Jeronimo—Apartado 3087, 3000-995 Coimbra, Portugal.
Email: allegretto70@gmail.com.

Agus Salim (Indonesia) is the Programme Manager in PATTIRO, a non-profit organization set up in 1999 to promote decentralization and good governance through transparency, social accountability for public services and public finance. Previously, he was a director in PATTIRO Banten. His work is focused on public services issues, including the promotion of gender responsive budgeting at national and sub-national levels. He was also involved in the formulation of technical guidelines for gender mainstreaming through gender responsive planning and budgeting which has been adopted by the national and sub-national government. He was involved with Open Government Partnership (OGP), a multilateral initiative to promote transparency, as PATTIRO's representative.

Address: Agus Salim, PATTIRO (Pusat Telaah dan Informasi Regional), Jl. Intan No. 81, Cilandak Barat Jakarta Selatan 12430, Indonesia.
Email: agus@pattiro.org.

Aloyah A. Bakar (Malaysia) is presently the Chief Executive Officer of the Penang Women's Development Corporation (PWDC), a state-linked company in Penang, Malaysia. As the former Project Director for Gender Responsive and Participatory

C. Ng (ed.), *Gender Responsive and Participatory Budgeting*,
SpringerBriefs in Environment, Security, Development and Peace 22,
DOI 10.1007/978-3-319-24496-9

Budgeting, her commitment and dedication in embarking on joint projects on gender responsive and participatory budgeting with the local councils and citizens on good governance and local democracy practices were rewarded when the project earned international recognition. The most significant was the highly competitive IOPD Distinction "Best Practice in Citizen Participation" where the Gender Responsive and Participatory Budgeting of Penang was recognized with a special mention by the International Observatory on Participatory Democracy (IOPD). She has diverse experiences over a span of 25 years from several private and multinational corporations, fine-tuning her many capacities and expertise in the areas of human resource management, entrepreneurial enterprises and non-governmental organisations.

Address: Aloyah Bakar, Penang Women's Development Corporation, Level 47, KOMTAR, 10503 Penang, Malaysia.
Email: aloyah.bakar@pwdc.org.my.

Roberto Falanga (Italy/Portugal) received his Master's degree from the University La Sapienza of Rome for his thesis on the psychological approaches to participatory urban planning. He was a Ph.D. fellow at the Centre for Social Studies (University of Coimbra) and member of the project OPtar, and received his Ph.D. with Summa cum Laude and Honours in 2013 for his study on cultural changes in public authorities implementing participatory processes. In 2014, he joined the Institute of Social Sciences (University of Lisbon) for the European FP7 Project "Mobilizing the Potential of Active Ageing in Europe" as postdoctoral fellow. He is also advisor for participatory programmes in Lisbon with the NGO In Loco and is Assistant Professor of Social Sciences at the Catholic University of Viseu, Portugal.

Address: Dr. Roberto Falanga, Postdoc Researcher, ICS (Institute of Social Sciences)—University of Lisbon, Av. Professor Aníbal de Bettencourt, 9 1600-189 Lisbon, Portugal.
Email: roberto.falanga@hotmail.it.

Regina Frey (Germany) is a Political Scientist and Gender Mainstreaming Expert. She has considerable experience in applied gender research projects (evaluation of gender equality strategies, programmes and projects), and has been a consultant for the implementation of gender mainstreaming and gender budgeting for public administrations, universities and NGOs. She also has a background in international development and participatory project planning. She is a Gender Trainer and Lecturer and has authored several publications on Gender Mainstreaming, Gender Budgeting, Gender Training, Theories of Gender and Managing Diversity. She founded the consulting firm genderbüro in Berlin in 2002.

Address: Dr. Regina Frey, genderbüro, Harzer Straße 37, D-12059 Berlin, Germany.
Email: frey@gender.de.

Sunny George (India) is HUDCO Chair Professor and Director of the South Asia School of Local Governance at the Kerala Institute of Local Administration (KILA), Kerala, India. His areas of interest include Applied Economics, Local Economic Development, Decentralization and Local Governance, Participatory Planning, Training and Capacity Building and Managing Development Projects. He conducts programmes for local governments in Kerala as well as for participants from other countries, including on Participatory Planning and Poverty Reduction. He has managed several development projects including major partnership projects with UNDP, the World Bank, and UN-Habitat. He has also been a consultant, for example on the project "Strengthening Local Government in Bangladesh" supported by CORDAID.

Address: Dr. Sunny George, Professor, Kerala Institute of Local Administration (KILA), Thrissur, Kerala 680 581, India.
Email: sunnygeorgekerala@gmail.com.

Elisabeth Klatzer (Austria) has extensive experience in research and civil society activism, and as a writer, lecturer and consultant in the fields of Gender Responsive Budgeting, Public Finance Management and Feminist Economics. She has experience in public administration and has been involved in the integration of GRB in the Austrian budget reform. She has worked in many countries, including Austria, Turkey, occupied Palestinian Territory, Bosnia and Herzegovina, Serbia, Ukraine, Tajikistan, and FYR Macedonia. She has recently finished a research project promoting Gender Responsive Budgeting and Gender Mainstreaming in cooperation with Ugandan universities. Elisabeth has a Ph.D. in Economics from Vienna University of Economics and Business, and a Master's in Public Administration from Harvard University.

Address: Dr. Elisabeth Klatzer, Hans Weigelgasse 21/2, A-2344 Maria Enzersdorf, Austria.
Email: elisabeth.klatzer@gmx.net.

Maimunah Mohd Sharif (Malaysia) is the President of the Municipal Council of Seberang Perai (MPSP) since March 2011. Her remarkable contribution to Penang and Seberang Perai has been recognized by the Penang State government, Malaysian government and international organizations through awards such as "Best City Manager" and "Best City" by Europe Business Assembly, Stresa, Italy and "Planner of The Year 2014" by the Malaysian Institute of Planners. She began her career as a Town Planning Officer in the Municipal Council of Penang Island (MPPP) in 1985, rising to become the Director, Department of Planning and Development, MPPP in 2003. In November 2009, she established the George Town World Heritage Incorporated to manage the George Town World Heritage Site. Maimunah has presented papers in seminars, forums and conferences, both in Malaysia and abroad.

Address: Maimunah Mohd Sharif, Majlis Perbandaran Seberang Perai, Jalan Perda Utama, Bandar Perda, 14000 Bukit Mertajam, Pulau Pinang, Malaysia.
Email: ydp@mpsp.gov.my.

Purusottam Nepal (Nepal) is Under Secretary of the Ministry of Federal Affairs and Local Development (MOFALD), working as National Programme Manager of Local Governance and Community Development Programme (LGCDP), a framework programme for improving inclusive local government in Nepal, funded by 14 Development Partners and the Government of Nepal. He has been involved in the planning and implementation of various programmes on a national scale to progressively restructure the country and solve problems arising from discrimination by gender, race, caste and religion.

Address: Purusottam Nepal, Under Secretary/National Programme Manager, Local Governance and Community Development Programme (LGCDP), Programme Support Unit, Ministry of Federal Affairs and Local Development, Singha Darbar, Kathmandu, Nepal.
Email: nepalpuru@yahoo.com.

Patahiyah Binti Ismail (Malaysia) was the first woman Municipal Secretary, Municipal Council of Penang Island (MPPP) in June 2009 and was subsequently appointed the first woman President in March 2010. In 2015, she became the first Mayor of the newly-formed Penang City Municipal Council. She began her career at MPPP in 1985 as an architect with the Building Department and was promoted to Director of the department in 1992. In July 2012, in recognition of her role in and continuous contribution towards the development of the state, His Excellency the Governor of Penang conferred upon her a state honorary award which carries the title Dato'. In 2013, she also received the "Best Municipal Manager" and "Best Cities" awards from the Europe Business Assembly based in Oxford, United Kingdom.

Address: Patahiyah Binti Ismail, Majlis Bandaran Pulau Pinang, Paras 4, KOMTAR, Jalan Penang, 10675 Pulau Pinang, Malaysia.
Email: Datukbandar@mbpp.gov.my.

Herculano (Allan) Ronolo (Philippines) is City Administrator, Malaybalay City, Philippines. He joined the government service in 1986 after teaching for nine years. He was appointed as the Local Planning and Development Coordinator and has served as the GAD Focal Person up to the present. He has been very involved in the conception and implementation of the Malaybalay Integrated Survey System (MISS). He has also been instrumental in the city consistently winning awards for being regionally outstanding in nutrition, child friendliness, best local development council, and best monitoring committee. His varied expertise and experience have taken him to discussions, conferences and planning teams across the Philippines and abroad.

Address: Herculano S. Ronolo, City Administrator's Office, City Hall, C. M. Recto St. Barangay 1 City of Malaybalay, Bukidnon 8700.
Email: allanronolo_56@yahoo.com.

Shariza Kamarudin (Malaysia) is Senior Project Officer for the GRPB Project being implemented by the Penang Women's Development Corporation in Penang, Malaysia. She has been with the GRPB team since February 2012 and has been particularly involved in community-based work with two low-cost flat communities. Shariza graduated from the University of Malaya with a BA in Literature and Gender Studies and has worked for a number of leading women's organizations in the country, including as Project/Programme Officer for the Women's Centre for Change (WCC), Penang, and for Sisters in Islam (SIS) and as a Research Assistant for the Women's Development Research Centre (KANITA) in Universiti Sains Malaysia, Penang.

Address: Shariza Kamarudin, No. 8 Pintas Semilang 1, Seberang Jaya, 13700 Seberang Perai, Pulau Pinang, Malaysia.
Email: shariza.kamarudin@pwdc.org.my.

About this Book

This unique book explores the exciting new democratic spaces that open up when budgets get participatory. The experiences of gender responsive budgeting (GRB) meet the world of participatory budgeting (PB), both of which have gained traction since the 1980s. The chapters bring together GRB and PB policy makers, practitioners, researchers and civil society actors to share and review their respective journeys. The book is a valuable contribution to current debates on re-thinking GRB and PB and the need to synergize both approaches to achieve gender equality, social justice and good governance. It is relevant to professionals in government and the private sector as well as to non-governmental organizations and students in gender and development and public policy.

"The book addresses the critical issue of citizen participation through a conversation between those involved in participatory budgeting and gender responsive budgeting approaches. It is rich in case studies and offers the promise of a new direction for entwining women's and men's participation by building a broad-based expertise in budget analysis and good financial governance" (Rhonda Sharp).

Contents

C. Ng (ed.), *Gender Responsive and Participatory Budgeting*,
SpringerBriefs in Environment, Security, Development and Peace 22,
DOI 10.1007/978-3-319-24496-9

Part II: Implementing Gender Responsive Budgeting Tools for Gender Equality and Social Justice.

Integrating Gender Equality, Women's Rights and Participation in the Budget Process: A Survey of Entry Points and Practical Examples: *Elisabeth Klatzer.*

Malaybalay City Integrated Survey System: A Tool for Gender Responsive Budgeting in Local Governance: *Herculano S. Ronolo.*

Part III: Institutionalizing GRB: Towards Better Accountability and Good Governance.

Localizing Gender and Participatory Budgeting: Challenges of Institutionalization in Penang, Malaysia: *Aloyah A. Bakar, Patahiyah Ismail and Maimunah Mohd Sharif.*

Institutionalizing Gender Responsive Budgeting in National and Local Governments in Nepal: *Purusottam Nepal.*

Gender Responsive Budgeting: State and Civil Society Initiatives at National and Sub-national Levels in Indonesia: *Agus Salim.*

Penang Women's Development Corporation (PWDC)—About the Editor—About the Authors—Index.

More on this book is at:
http://www.afes-press-books.de/html/SpringerBriefs_ESDP_22.htm.

Index

C. Ng (ed.), *Gender Responsive and Participatory Budgeting*,
SpringerBriefs in Environment, Security, Development and Peace 22,
DOI 10.1007/978-3-319-24496-9

Printed by Printforce, United Kingdom